Extra-Terrestrial Liberty

Extra-Terrestrial Liberty

An enquiry into
the nature and causes of
tyrannical government
beyond the Earth

Charles S Cockell

Shoving Leopard

First published in 2013 by

Shoving Leopard
Flat 2F3, 8 Edina Street
Edinburgh
EH7 5PN
United Kingdom
http://www.shovingleopard.com/

Cover: Nial Smith Design, 13/4 Annandale Street,
Edinburgh, EH7 4AW

Text © Charles S. Cockell 2013

The right of the contributors to be identified as the authors of this work has been asserted in accordance with the Copyright, Designs and Patents Act 1998.

ISBN 978-1-905565-22-1

A catalogue record for this book is available from the British Library.

To would-be extra-terrestrial tyrants

Acknowledgements

The essays in this book are a collection of essays published in the Journal of the British Interplanetary Society, but which together constitute a single line of thinking.

I am grateful to many people who have been responsible for causing these ideas to take root, particularly the friends and colleagues I have enjoyed speaking to in my lifelong interest in the establishment of a permanent human presence beyond the Earth. They are too numerous to list.

It was a chance encounter with a copy of Rousseau's *The Social Contract*, in a second-hand bookshop in Cambridge in 2001, that ignited my excitement and interest in political philosophy and inevitably led to a link with space exploration and settlement. Rousseau has been responsible for a lot, but contributing to ideas about how we might settle space probably would have surprised even him.

I'd like to thank Janet de Vigne for publishing the essays as a collection.

Contents

Acknowledgements   *vii*
Preface   *xi*

Extra-Terrestrial Liberty   1
   1. Introduction   5
   2. Liberty and the Response to Tyranny   9
   3. Extra-Terrestrial Environments and their Differences,
      Physical and Social, to the Earth   13
   4. Influence of Extra-Terrestrial Conditions On Society   17
   5. The Effect of Extra-Terrestrial Environments on
      the Buffers to Tyranny   21
   6. Extra-Terrestrial Liberty   25
   7. The Influence of Population Size   31
   8. The Bearing of the Political and Economic System   34
   9. The Character of Extra-Terrestrial Freedom   40
   10. Children and Extra-Terrestrial Freedom   44
   11. Religion and Extra-Terrestrial Freedom   47
   12. Private Property   51
   13. The Tendency to Tyrannise   55
   14. The Liberation of Creativity   58
   Notes and references   61

Liberty and the Limits to the Extra-Terrestrial State 77
   1. Introduction 81
   2. The Justification for the Extra-Terrestrial State 84
   3. The Problem of Oxygen 88
   4. The Transition from a Centrally Planned to a
        Maximally Free Extra-Terrestrial Economy 94
   5. Education 103
   6. The Culture of Liberty in Society 108
   7. Religious Institutions 112
   8. Private Land and Property 115
   9. The Limits and Purpose of the Extra-Terrestrial State 118
   Notes and References 125

Liberty Across Light Years – The Freedom of Future Space Settlers Compared to that of the Ancients and the Moderns 145

On Tyranny 149

The Causes and Consequences of Extra-Terrestrial Tyranny 151
   1. Introduction 153
   2. The Consequences of Extra-Terrestrial Tyranny 156
   3. What is Extra-Terrestrial Tyranny? 160
   4. The Historical Origins of Tyranny 167
   5. The Environmental Origins of Tyranny 171
   6. The Causes of Economic Tyranny 179
   7. The Problem with Lunar Leninism and Martian Marxism 187
   8. Tyranny and the New Society 194
   9. Conclusions 203
   Notes and References 206

Other books by Charles Cockell *231*

Preface

An enquiry into the nature of freedom and tyranny in outer space may seem distant to the people of the Earth. It is speculative and apparently lacking in any immediacy to the lives and daily experiences of most individuals.

But the study of liberty in outer space will undoubtedly have enormous importance for those who eventually live on the planets and moons of the Solar Systems and the vast realms of its empty spaces. As the discussion on terrestrial liberty has fuelled debates that have raged since Locke, Hobbes and Rousseau brought them to public attention, so any literature of extra-terrestrial liberty will be eagerly seized by later generations in the hope that they might glean something of use of to their own societies in space - for determining what institutions they should create and how the conditions for liberty are to be maximised.

No institution established either to defend liberty or that has a character or purpose that contributes to the promulgation of liberty can expect, or more to the point, should out of prudence, expect eternal longevity. Individuals and institutions are constantly exploring the conditions for freedom in their given time and place and seeking to find new ways to advance it, rearranging institutions and contemplating new mechanisms to protect it from decay. Complacency becomes, very rapidly, the worst enemy of liberty. This will be as true in space as it is on the Earth. There is no magical insight into extra-terrestrial liberty. As any given extra-terrestrial society evolves from a small outpost into a growing settlement and eventually into something akin to a nation, the conditions for liberty will change and with it the nature of institutions required to make it flourish. However, the general environmental conditions within which these transformations will occur and the character of the human personality will not change. Any contribution to the discussion on the form that freedom might take in outer space is therefore an invaluable addition to the pantheon of literature and ideas from which individuals can draw to shape and reshape the institutions to ensure that their societies do not succumb to despotism.

The denizens of the extra-terrestrial frontier will be confronted by a myriad of questions that will define the core of their political philosophy: Can a person be free when the oxygen they breathe is the result of a manufacturing process controlled by someone else? Will the interdependence required to survive in the extremities of the extra-terrestrial environment destroy individualism? What are the obligations of the individual to the extra-terrestrial state? How can we talk of extra-terrestrial liberty when everyone is imprisoned?

Even the most extreme terrestrial environments, such as the polar regions, cannot provide an adequate insight into how freedom will develop, since the combination of extremes found in these places is different from those experienced beyond the Earth. At least societies that persistent in extreme conditions, such as the Inuit of the Arctic, possess the freedom of breathing freely available air. This is a luxury denied to people in any extra-terrestrial environment.

The reader may already have noticed that not all commodities that we require for survival on the Earth, such as water, can be availed free of state intervention. Water is provided through water treatment works which are subject to strict state water safety standards, and in that respect we are all subjugated to these controls. However, none of these requirements, should they be denied to us, threaten instant death in the same way that shutting down air supplies in space does. In many countries on the Earth, to attempt to close down all water supplies would give people enough time to seek out water from rivers and other natural sources whilst launching a revolution against their government. Denying people water, except amongst people in deserts, is not a fool proof means of coercion.

To understand why the lack of readily available oxygen should be so relevant for an enquiry into extra-terrestrial liberty it is only necessary to consider a simple analogy. There are many ways to think about asteroids, the Moon, Mars and the free spaces between them, the lack of breathable atmospheres and the conditions that they present to human beings. But I think that an accurate, if not rather crude and insensitive, way in which to picture these environments is as vast gas chambers. Land on them without a spacesuit and you will be instantly asphyxiated. So to live within an outpost or settlement in any of these locations makes a person daily in debt to the state, or the corporations that run the settlement, for the gift of oxygen. It is to officials that you owe your gratitude for their generosity and kindness in preventing you from being naturally gassed, even if you have paid them. To take this analogy to a more perverse level, one might consider it to be an inverted Holocaust. The victims of the Holocaust were free to breathe air, but the Nazis took it from them by gassing them. In space, the default, natural position is to be gassed; in providing oxygen, the state or corporation shows you benevolence.

Now one does not have to believe that governments or corporations need to be overtly aware of the fact that they have corralled under them a group of people protected from a natural gas chamber to derive from this analogy some insight into what the environment will do to political power. Any dissent, criticism, or attempted overthrow of those in power can be met with an apology for the excessive coercion and tyranny under which people live and then the offer for people who wish to free themselves to leave the settlement. Maybe the government will be so gracious as to open the airlock for the liberated dissenter? When the outside environment promises instant death, the boundaries of tyranny within an extra-terrestrial settlement have no limits. It is to this, the problem of extra-terrestrial liberty, that I turn again in these essays.

Despite these very different conditions to the terrestrial case, all of the questions that inhabitants of space will ask themselves fold together into a single question that has concerned people on the Earth since the birth of human societies: Does the spread of civilization, in this case into the extra-terrestrial environment, represent an opportunity for the expansion of liberty or an opportunity for the expansion of tyranny? At is on the Earth, it is neither one nor the other, but it is finding the balance between freedom and control that will present a challenge of immense proportions for societies that develop in parlous physical conditions that are inherently prone to tyranny. The opportunity that the airless, radiation bombarded locations of the Universe offer for dictatorship and the weight of repression and control that despotism will exert on the structures of freedom to the point of their complete collapse will force upon any free-thinking people in outer space a requirement for a deep and continual discussion on their liberty. For them, more than any society that has existed on the Earth, political philosophy will not merely be an enticing intellectual exercise, but a crucial instrument in the construction and maintenance of free minds.

A present-day enquiry into the causes of tyrannical government beyond the Earth may be speculation, and it may be founded on terrestrial experiences that cannot truly approximate to extra-terrestrial conditions, but it is a start. It provides a position from which the eventual reality of living in space can be used to advance a realistic understanding; it yields a foundation from which to formulate an accurate view of the conditions for liberty beyond the Earth when those settlements are eventually established.

There is likely to be much more use for this enquiry than merely developing the apparatus of freedom in outer space. It is likely that the future of liberty on the Earth will depend on the character of freedom in space. In the short-term, a few scattered tyrannies in space can offer little threat to terrestrial liberty. More consumed with maintaining their own existence, and probably, for a good deal of time, dependent on resources from the Earth, they will work within the sphere of influence of the Earth and they will not possess the wherewithal to challenge terrestrial institutions.

However, the Earth sits at the bottom of a gravity well and outer space offers an energetically and psychologically advantageous position from which to look down on it. This fact has been recognised by military strategists since the invention of the rocket. It has been one of the driving urges to prevent the construction of weapons in space, a tacit understanding that it is better for no-one to have military control of space than one person to establish themselves as the master of the strategically superior position of Earth orbits and beyond.

The emergence of tyranny in outer space does not require the spread of weapons into space to threaten the Earth. The mere existence of tyrants peering from the top of the gravity well down to the Earth, with its free people watching up at the ogre-like stares of dictators will be enough to weaken the resolve of liberty on the Earth. Commanding the resources of outer space and means of supply of minerals of importance to Earth industries, these tyrants could eventually wield economic influence on the political institutions of terrestrial states.

If liberty should be extinguished in outer space it is likely that eventually the Earth too will be eclipsed by a dark, cold shadow of tyranny. It seems to me that those people who control outer space will eventually control human freedom.

It is no surprise, then, that this book is not dedicated to anyone I know or have known or even to future people of the Earth. It is dedicated to the free-thinking people of outer space, to those who have yet to be born. It is dedicated to the workers, tourists, families and individuals on the Moon, Mars and in the spaces in between, to the lone worker tending to a robotic plant on an asteroid.

As with any discussion on liberty, this book is, by virtue of its subject matter, a contribution to revolutionary literature. It should be read by those, who, at the end of their tireless patience and tolerance for extra-terrestrial authorities that use safety and the threat of the lethal conditions of space as an excuse for draconian systems of control and repression, wish to construct a new order. Of course, the specific mechanisms by which attacks on liberty are instituted in any particular place and in any specific time cannot be predicted here, but the reasons why they come about, I contend, are understandable. There is no good excuse for it. Although the conditions of space will inexorably drive societies towards tyrannous states of organisation on account of the dangers they face, it makes the responsibility of authorities to pursue the tenets of liberty even more acute, their duty to seek the views and ideas of the people under their charge as to how to achieve the route to freedom even more urgent. Authorities that fail in their wider duty to contribute toward the human effort to hold despotism at bay in extra-terrestrial environments, and abnegate their responsibility to prevent it spreading to other settlements and thence to the Earth, forfeit their right to continue in power, and the people they rule should assume their right to seize the means of production and supply through all avenues at their disposal, including revolution.

In this set of essays I do not attempt to provide some type of complete insight or investigation of extra-terrestrial freedom. The purpose of my discussion is to contribute just one set of ideas to the long journey of constructing the intellectual foundations of extra-terrestrial liberty by drawing, as best we can, from the lessons we have learned on the Earth.

Charles S. Cockell
February 2013

Extra-Terrestrial Liberty

I wrote this essay out of surprise. I was intrigued that after more than three hundred years of political philosophy, a tradition of discourse on liberty reaching back to Hobbes and Locke (although I should say itself inspired by ancient Greek and Roman ideas) there was not a single essay that dealt explicitly with extra-terrestrial liberty. Of course, many authors had previously discussed societies in space and many science fiction writers had touched on the how communities might develop in space and what the social conditions of those branches of civilisation might be. However, the subject of extra-terrestrial liberty had remained untouched as an academic discussion drawing directly on the long tradition of political philosophy applied to the Earth. This essay was an attempt to contribute to a direct discussion on what freedom might be beyond the Earth and how it might compare to classical Enlightenment concepts of liberty.

Extra-Terrestrial Liberty

The lethal environmental conditions in outer space and the surfaces of other planetary bodies will force a need for regulations to maintain safety to an extent hitherto not seen on the Earth, even in polar environments. The level of inter-dependence between individuals that will emerge will provide mechanisms for exerting substantial control. In extra-terrestrial environments, traditional buffers to tyranny that exist on the Earth are either absent or much weaker. Legislative and political mechanisms used to protect freedom will be needed to such a degree that they themselves are likely to become a form of despotism. Thus, the most profound irony of the settlement of space is that the endless and apparently free expanses of interplanetary and interstellar space will in fact allow for, and nurture, some of the most appalling tyrannies that human society can contrive. Thwarting this tyranny will be the greatest social challenge in the successful establishment of extra-terrestrial settlements.

1. Introduction

'What extent of liberty is possible in extra-terrestrial environments?', 'Can outer space host free worlds?' As people begin to establish a permanent presence in space, from the smallest settlements to the largest self-sustaining societies, they will ask these questions and thereby generate new discussion of the social and legal definitions of 'liberty' and 'freedom'. The tradition of social and political thinking on Earth, from early Greek thinkers, particularly Plato [1], to more contemporary figures such as Berlin will guide some of these ideas. But I will argue here that extra-terrestrial environments will necessarily result in less freedom and more tyrannical oversight of human lives.

It is beyond the scope of this essay to suggest a civic structure. However, if it is the case that extra-terrestrial environments will be places where liberty will less easily flourish than on the Earth and where it is likely that individual and civil freedom will be more fragile, then an important task will be to find ways to encourage and nurture the conditions for liberty [2]. It is first necessary, therefore, to understand and define the characteristics of liberty, and how it is most likely to be expressed in extra-terrestrial environments. It then becomes possible to identify those types of institutions that need to be strengthened or created to encourage freedom and those institutions that should be kept subordinate, or even never created, through which tyranny is most likely to be expressed [3]. From this secondary analysis it is then possible to define the architecture of the extra-terrestrial state.

To begin with, it is useful to identify some basic definitions and precepts with respect to the whole notion of freedom. It is worth recalling that there is no such entity as a 'free society'. It has become something of a cliché that this phrase is a contradiction in terms. To have a society implies the existence of institutions and organisations through which people are organised into a 'society'. The existence of institutions and organisations implies rules, conventions and laws through which those institutions and their constituent members are organised. The existence of rules, conventions and laws implies that individuals cannot be completely free, regardless of whether those regulations are imposed by dictate or even democratic consent. Thus, a society implies restraints on freedoms and it follows that there is no such absolute entity as a 'free society' [4]. The discussion at hand is therefore not whether extra-terrestrial environments can host 'free societies', but, as it is on Earth, about what extent of freedom is possible and in which realms of human activities liberty can be most easily promulgated.

A practical example of this idea, to which I will return later, because it is essential to an analysis of the nature of extra-terrestrial liberty, is freedom of movement. The ability to move unhindered, from and to one's house for example, is considered a fundamental type of freedom. In societies considered despotic, freedom of movement is often restricted. state passport controls can prevent individuals, travelling, of their own accord, to other countries. However, internal national constraints on movement may additionally be implemented by the state. This restriction of freedom is often actually regarded as one of the *defining* characteristics of a society considered to be totalitarian.

In the freest societies there is not unrestricted freedom of movement, however. Military establishments are usually out of bounds to the general public, unless individuals have the requisite paperwork. Other government offices, buildings and land usually require specific types of passes. Even private land is not freely accessible to all people, who can become subject to the laws on trespass if they violate regions that are legally recognised to be the private property of others. There is therefore a gradation of freedom of movement on different types of land and property on which both government and private individuals can legally claim control over the movements of others. There is no absolute freedom of movement associated with a free society, but there is increasing freedom of movement in an increasing diversity of regions in those societies considered to be freer. Similar observations apply to the idea of the 'free press' and public access to information. Neither of these concepts expresses an absolute end point that is found in any real society on Earth.

It is also mandatory to make some attempt to understand what is meant by the very word 'freedom'. The word is used glibly, by politicians and governments alike, as a rallying call to domestic policy and, in particular, foreign policies. It is an effective means of implying that those who disagree with your policies are, by default, against 'freedom'. In this context its political use can be mischievous and highly insidious. In addition to the problem of the loose use of the word, freedom is not an isolated property of individual or group behaviour: the freedom of one individual or group to act often implies a freedom infringed elsewhere. A person who expresses their freedom to expand the size of their seaside home may infringe the freedom that another individual cherishes to view the sea unhindered from their home. This situation, 'the paradox of freedom' as it is sometimes called [5], encompasses the problem that allowing individuals to do one thing, performing actions protected either by the law or culture of a society, may result in situations that others consider an infringement of the actions and ideas that they would prefer to express, and which they consider to be representative of their freedom.

Despite these problems, it is also wise not to stray into the error of implying that there is no such thing as 'freedom' or that any form of society is merely a variation on the expression of liberty. Some societies do uphold the liberty of individuals in as many activities as possible, in a very real sense - as a cherished goal. Overtly totalitarian governments can be distinguished from governments that pursue the general goal of freedom, although I shall later argue that the structures of such societies are not categorically different. In this essay, the concept of freedom is used in a general way to mean the extent to which individuals consider themselves both free of external interference in a large number of their daily activities and thoughts, and free to implement their own ideas and plans. In other words I am referring to a social conception of freedom. The sense of freedom used here is also distinct from the sense it which it means emancipation from determinism or instinct [6], a question important to people during the Enlightenment. It is also distinct from attempts to escape the imposition of extreme environments. In the sense that I shall argue here that people in extra-terrestrial societies will be very much under the tyranny of Nature and the 'irrational' whims of extra-terrestrial environmental conditions, one could make the case that extra-terrestrial freedom will be an attempt to bring order and rationality into an otherwise precarious existence under the direction of powerful natural forces.

The pursuit of individual liberty and the conflict it can cause in attempts to advance the collective interests of human society has been one of the greatest, and most fascinating, challenges to the assembly and maintenance of human societies. During the twentieth century, technological advances in air power and other forms of warfare made the century the most destructive in the history of human society; collective visions of human progress and those favouring more emphasis on the pursuit of individual liberty came head to head in ways previously not possible [7].

Such is the importance of the resolution of these matters to human lives that it seems inevitable that equally strong convictions and points of view will be associated with these questions in outer space. It is my objective here to explore the nature of extra-terrestrial liberty.

2. Liberty and the Response to Tyranny

Before examining how liberty might be expressed or eroded in extra-terrestrial environments, it is first useful to understand more generally what happens to people when they are subject to increasing control in an increasing range of their daily activities. These basic precepts can then be applied to extra-terrestrial environments, in order to predict the type of society that might emerge, and how individuals are likely to respond to it. For the sake of convenience, we can recognise four responses to tyranny that are open to individuals or society collectively.

As a government or organisation makes ever-greater incursions into the activities of human individuals, these individuals may acquiesce and accept these interferences. As the intrusions increase in number and scope, the population may continue to accept them until they become completely and willingly enslaved. Provided that they think that there is no alternative, this will be the path of least resistance: it will lead to a malleable population that will respond to whatever demands are placed upon it. Such acceptance of rules may come from within the individual, or, if the governing powers are politically adept, they may be able to convince individuals that such incursions into their lives and liberties are in their own interest, and therefore represent a type of liberty. Later I will show how extra-terrestrial environments make this possible.

Assuming that the population continue to accept their rulers, then liberty will be confined into an ever-smaller sphere of activities, until it is almost extinguished. In some sense, this is hardly a 'buffer' against tyranny as it is a reluctant compliance; but, on the other hand, it does represent a possible response to tyranny as a means to buffer against social disintegration and disorder, which individuals may perceive as the alternative choice, or it may be presented to them as the alternative. This response will ultimately fail in the sense that authorities can become so tyrannous that the population eventually become vassals.

Many may find the realism of such an agreeable response to tyranny questionable, but such complicity has precedents in history. People have immense incentives to maintain the status quo. Not only is it beneficial for the stability of their own lives, but also for their children and family. Many people would prefer stable tyranny in which they can raise their families and live predictable lives, rather than risk the instability caused by a social uprising against their government until, that is, the tyranny becomes so bad that instability seems a small price to remove it. Reluctant complicity to new rules and regulations is often the first stage of a social response to increasing incursions on individual liberty. How can individuals then escape the fate of increasing repression?

There are three further courses of action, or responses, which do not involve complete acquiescence. The least rebellious of these is to escape into oneself—a Stoic retreat into one's own thoughts, which remain beyond the reach of laws and regulations, and outside the control of the state. This is a common course of action for those who feel repressed by a society—to remove themselves from as many activities as possible that involve state oversight, and to fill their time in their own thoughts, over which they have complete jurisdiction. But, as Hegel recognised, usually this approach is ultimately doomed to fail, since no mind can completely disconnect itself from the world [8]. The thoughts one has necessarily require interaction with social organisations and other individuals. Stoic disconnection from society is only possible to a certain degree, and that that certain degree is determined by the individual, and especially by their strength of character and mental self-sufficiency.

Once their Stoic retreat has failed, the next approach open to individuals is to rebel openly, and to disagree with the rules and regulations. To pursue this path, individuals require mechanisms for organising. It is reasonable to assume that, in most societies, a single individual, acting alone, has a limited capacity to halt the encroachment of tyranny, because the invasion of the state into many individual lives usually implies an organisational force against which it is difficult for a single individual to fight. Therefore, some ability to collectively organise people must be available. This may take the form of physical organisation (gatherings, meetings, rallies), or it may involve the organisation of individuals through electronic communication.

If the voice of the group of dissenters is loud enough or the threat to the existing system great enough, then they may effect change and halt the spread of a tyrannical influence. The extent to which they are successful depends, of course, on the level to which the individuals and organisations imposing control are determined to maintain their position, implement their ideas or repress dissent. Even on Earth, in some communist societies such determination was sufficient to completely repress political dissent of most kinds, and made this third avenue of resistance to tyranny not generally available.

In a society that allows dissent, or even encourages it, this avenue is the most effective means to hold at bay or reverse state practices. This is, of course, the principle of the democratic polity, in which elections, referenda and similar mechanisms are the devices through which people 'dissent' from the prevailing direction of society and set it on a new course, whether in small affairs of state or in the very leaders they choose to govern them.

The fourth course, following the failure or limited success of a Stoic retreat or direct action, is to escape altogether physically. However, in a society that imposes border controls this approach too may fail. Efforts to escape a government can succeed in relatively free societies, where cross-border movement is possible, and where some individuals feel sufficiently infringed in their freedom to move elsewhere; and in societies that suffer an invasion from outside, in the form of occupying armies. In such circumstances, individuals may seek to escape into forests or uninhabited regions, and await (hopefully) a restoration of freedom through military liberation. Whether these are permanent solutions depends on many diverse social factors such as, for instance, whether they have left behind family or friends whom they may wish to see again, and the likelihood of achieving refugee status in another country, assuming they manage to reach one.

Each of these courses of action in turn involves an increasingly extreme approach, both in mental effort and in physical requirements. In highly repressive societies all four paths may be impractical. The first approach, acquiescence, may eventually be negated by an individual's refusal to be subjugated, the eventual enslavement of the population and te destruction of most freedome. The second, Stoic introversion, will fail because of the ultimate need for human contact, which forces the individual to interact with society. The third may be rendered useless by political mechanisms of control over individuals and their ability to organise, which prevent overt dissent. The final course may be rendered impossible by tight border controls, or the lack of places to which individuals can move. If all four of these responses to tyranny fail, then absolute tyranny becomes possible. I use the phrase 'absolute tyranny' specifically to mean the potential for tyranny in a society in which the buffers to tyranny have failed, or have been rendered ineffective by those running the society - or by the environment in which the society operates, which is most pertinent to the discussion here.

3. Extra-Terrestrial Environments and their Differences, Physical and Social, to the Earth

All extra-terrestrial environments that are currently known and might be capable of hosting a permanent human presence (I mean primarily outer space, the moon and Mars in the short term, but eventually possibly also places in the outer solar system and beyond) have environmental extremes that are far worse than the most extreme environments on Earth [9]. The surface of Mars, for example, has a mean annual temperature of -60°C, and a higher background level of both ultraviolet and cosmic radiation compared to the Earth. Dust storms can rage across its surface for a third of the Martian year. These types of conditions place overwhelming technological requirements on humans to enable them to survive in these environments.

Consider the case of Mars. Drinking water can be obtained, but to do so it is necessary to identify, mine, melt, clean of dust and prepare in a form suitable for human consumption a source of ice, either permafrost in the ground or water ice at the poles. We know of no place on Mars where liquid water can be drunk directly from the surface. The Martian atmosphere contains water, which could be extracted, although the scale of the extraction systems required to collect sufficient liquid from an atmosphere with a mere 0.03% by volume of water are quite formidable. The availability of liquid water would be insuperably bound to an involved manufacturing process.

Food can be grown on Mars, but it requires a complexity of process. There is less natural light on Mars than on the Earth, because Mars is farther from the Sun, but there is sufficient to grow crops. However, the lack of an ozone shield on Mars means that the ultraviolet radiation at its surface is about a thousand times more damaging to biological material than that on the Earth. Standard window glass cuts out ultraviolet radiation and allows the penetration of light for photosynthesis, so this problem could be overcome. However, no crops can grow at the ambient atmospheric pressure on Mars, which is some one hundred times less than on the Earth; thus a 'greenhouse' must be pressurised. This requires, essentially, a light transmitting pressure vessel. This is not a trivial technology. The irrigation of the crops is itself linked into the water production system, creating a two-tier technological challenge that lies between agriculture and the consumption of food. The surface will require conditioning. There is no organic matter to yield a soil suitable for crops. To achieve high agricultural productivity, fertilizers and other conditioning compounds will probably be required.

The production of oxygen to breathe follows either from the water production system or from the atmospheric extraction apparatus. Once water is acquired, either from the ground or from the air, it can be electrolysed to produce oxygen (the hydrogen produced as a by-product might be combined in the Sabatier process with atmospheric carbon dioxide to yield methane fuel)[10]. Alternatively, the abundant carbon dioxide in the atmosphere could be cracked to yield oxygen (even this, however, might seem banal compared to the moon, where oxygen will probably have to be extracted from rocks).

There is no great value here in continuing a protracted discussion of how to live off the land on other planetary bodies, but this cursory description serves to emphasis the vast interconnected technical complex that would be required to supply even the most basic human needs.

One could argue that all extreme environments on Earth place technological requirements on people. If I walk out into the Antarctic winter without clothes I will die of hypothermia, eventually. In that sense, the Antarctic environment is inherently lethal [11]. However, the threshold to overcome in these extreme conditions is much lower than those needed to persist in extra-terrestrial environments. In addition, in many locations on the Earth it is possible to live physically independent of other people. For instance, in tropical forests, air, food and water can be gathered from natural sources. These basic commodities of human existence are in abundant natural supply free of technological syntheses and they are, in some locations of the world, for all practical purposes free of state control.

There have been past attempts to draw parallels between environments on Earth and in space. In a 2002 paper in *Interdisciplinary Science Reviews* entitled 'Mars is an awful place to live' [12], I argued that the Earth's polar regions bore most similarity to extra-terrestrial environments in terms of extrapolating a trajectory of social development, specifically on Mars. I also suggested that even if we did adopt that model, the population of Mars could rise to no more than 3 million people. But in almost all respects, apart from the vicious storms, the polar regions of the Earth are an improvement on the environments of places such as the moon and Mars; any attempt to draw social parallels between these terrestrial environments and the behaviour of people within them is likely to fall short of the reality of extra-terrestrial environments. Despite the limited source of experience on which we can draw to make truly trustworthy predictions about the nature of extra-terrestrial societies, the Earth's polar environments at least offer us some insights into the types of restrictions that we might expect the physical environment to cause.

All extra-terrestrial environments share one very fundamental difference with the Earth: anywhere on Earth one is free to breathe the air. In alien environments air is not a free commodity, acquired independently of technology: it will have to be manufactured. I contend that it is from this single crucial difference that many of the social differences between societies on Earth and those in space will derive.

The canonical comparison with the extra-terrestrial environment is the situation in polar stations on Earth. Where one can walk outside is strictly regulated for safety reasons. The use of food, water and other consumables is overseen and monitored, to ensure that demand does not outstrip supply. As the station grows larger, supplies may improve in scope and quantity, hence relieving some of the control over individual choice, but the environment outside does not change. No amount of social engineering or technology can change the fact that movement is restricted. At least in a polar station, when the weather is good, one can, completely independently, open the door, walk outside and escape society. The distance one can go, or would want to go, may be limited, but one can, for a brief period, escape with one's own thoughts. Now consider the case of an extra-terrestrial environment, where the atmosphere is instantaneously lethal. Even if the society took a relaxed attitude to outside movement and adopted a live and let live approach—allowing people to take spacesuits and wander freely outside whenever they chose—few individuals would do this without being thoroughly confident that their spacesuit was operational. Even the most cursory checks on a suit to ensure it pressurises properly place a restriction on outside movement that is repressive in comparison to that in many extreme environments on Earth.

This same pressure of restriction, and of interdependence between people, will apply to many more of people's day-to-day requirements. Each of the other extremes of extra-terrestrial environments—low temperature, high radiation, lack of liquid water, potentially toxic soils, lack of indigenous food—will impose restrictions on movement, and on expressions of freedom, that will radically influence the nature of extra-terrestrial society.

The failure of any systems that deliver vital services that allow survival in these extremes adds yet another layer of uncertainty, and thereby control. If a water recycling system fails in an Antarctic station, the inhabitants can continue to subsist on what they have in reserve, or they can, with sufficient motivation and improvisation, use energy sources to melt ice. Although such a situation may be dire, it is rarely lethal for the occupants. In fact, in most extreme environments on the Earth, technological failures that threaten to kill people instantly are rare: one notable exception, also common to extra-terrestrial habitats, is the possibility of fire in enclosed living spaces. In the extra-terrestrial case, a diversity of situations can, within a very short space of time, threaten impending death. The most severe examples are the failure of air or oxygen supply units, and the depressurisation of habitats. However, even the failure of water supply units is more dangerous than on the Earth. Although reserves can be stocked for emergency use, it is obvious that, unlike Antarctic ice, and even at the Martian polar ice caps where there is abundant water ice, it is less trivial for people to be able to walk out of a station and collect copious quantities.

4. Influence of Extra-Terrestrial Conditions On Society

We can make at least one prediction: to live permanently in extra-terrestrial environments will require much greater interdependence amongst people than in any environment on the Earth. The survival of every individual is dependent upon technology that is maintained by others, to an extent not experienced on Earth.

To a certain degree, the environment of any place affects the development of group customs and the behaviour of populations: for example, warm climates tend to create societies, such as continental European societies, where outdoor entertainment and living is more possible than in colder, more northern climates [13].

However, as the number of influences and their extremities increase, so the need to confront them through common means and through a well coordinated group discipline strengthens the collective psychology, and dulls the independence of mind expressed by individuals. This is readily observed amongst isolated populations in extreme conditions, such as mining stations, submarines and polar environments.

In extra-terrestrial conditions, the fear of disapprobation in a society in which interdependence amongst people is at the core of survival will create a population in which the majority of people are rarely willing to challenge the status quo, although it may have completely the opposite effect on some small part of the population.

The other physical aspects of the environment will aid this dulling of human individuality. In any extra-terrestrial environment one cares to consider, there is little variety, in comparison to the Earth. The moon is visually a grey wasteland. Mars is an improvement, with soaring mountains, white polar landscapes and canyon systems so immense they would engulf even the mightiest canyons on Earth; but this is a landscape dominated by reds, oranges and browns. Rarely, except in artificially created environments, will people naturally see greens and blues. As it is necessary to wear spacesuits, people will smell nothing but artificial odours. Perhaps the whiff of Martian dust, like the gunpowder smell of lunar dust reported by the Apollo astronauts, will sometimes enrich the odours of habitats and humans. With no leaves of plants through which wind can rustle, no rivers to trickle over rocks and no wildlife to fill the air with their calls and cries, the environment is an auditory wasteland. The long-term effects of these visual, olfactory and auditory privations on the human psychology can for now only be guessed at, but such sensory restriction cannot be beneficial to human mental health and the appreciation of the diversity of experience. Humans may inexorably become the servant of an insipid conformist outlook and, ultimately, of tyranny.

In a later section, I shall explore in more detail how this type of environment can influence the way in which liberty is expressed, but here is it useful to point out a basic principle. Once an environment becomes extreme and the society is made more interdependent, there are two fundamental ways in which the population and the collective customs of their society can respond. They can embrace these extremes as an unassailable reality, and learn to live within them or they can take on the environment as a foe. On Earth, populations that live in extreme environments, such as the Inuit of the High Arctic or the aborigines of the Australian outback, essentially assume the former posture. They have learnt through thousands of years of experience that fighting the local environment is futile, and that a more successful path to survival is to comply with the environment. By the word 'comply' I do not mean giving in, for that would result in death. I mean that they learn to nurture a culture and customs that fit within the environmental reality, such as hunting in the summer and storing food for the cold polar winter, in the case of the Inuit. Or they may build igloos—using the resources of the extreme environment to enhance their prospects of success. This may appear to differ very little from someone who 'fights' an environment, but there is an important psychological distinction between a people that sees their environment as something that should be respected and seek ways to live effectively within it, and a people who see an environment as something hostile to be overcome. I can think of no better analogy for this than Scott and Amundsen's race for the South Pole. Amundsen, as a Scandinavian, went for the Pole with dogs and used the methods learnt in the Arctic over many centuries. Scott confronted the Pole with tractors, ponies and every other technological innovation he could throw at it. In the former case, we see an attitude more reminiscent of attempting to live within the reality of the polar environment and employing approaches compatible with it; in the latter case, we see a psychology that views the polar environment as something to be defeated [14].

There can be no doubt that extra-terrestrial environments will, because they are so extreme, require a fight for survival, but there is also likely to be a similar dichotomy in how these environments are viewed: either as an unconquerable challenge that demands respect and an approach that maximises the extent to which the lives of people are moulded around the environment, or as an enemy to be overwhelmed by the application of will power and technology.

It is not easy to predict which attitude people will adopt in different extra-terrestrial environments, but we can assume that either response demands an essentially more collective approach to social organisation than is necessary in more benign environments. In the former case, people are brought together by customs, regulations and behaviour that are considered appropriate in addressing the need to assimilate into the environment, and they treat it in the way required for survival. In the latter case, the vision of the 'fight' against an environmental foe will similarly encourage a collective response, a war if you will, against a common enemy [15]. The difference between these two responses is, it seems, that the latter is inclined to encourage a stronger argument for central control of individual lives than the former. In the former case, provided some minimal set of behaviours is adhered to, to ensure that society and individuals are safe from the errors of individuals, then liberties that do not impinge upon these requirements are more plausibly allowable. However, the war against the environment impels people to adopt an attitude that requires more mobilisation, and mobilisation lends itself to tyrannical behaviour by dint of the fact that the much greater threat the environment represents provides a reason for issuing and promulgating dictates that align the population against the ever-present impending environmentally-caused destruction of society. A successful extra-terrestrial society, as least successful in the pursuit of liberal or even libertarian principles, will presumably adopt the former approach.

5. The Effect of Extra-Terrestrial Environments on the Buffers to Tyranny

With this in mind, I now focus on the effects of the physical environment and its influence on the responses to tyranny explicated earlier.

Consider the first response—acquiescence to imposed rules and regulations. Societies in which there is a strong collective psychology imposed by extreme environmental conditions, against which society must work to survive, imbue a greater willingness in individuals to acquiesce to rules and regulations. This tendency is because people are concerned for their safety and are more willing to give the benefit of the doubt to new rules that may be promoted and implemented under the banner of enhancing safety. Examples of such regulations include laws on the certification and licensing of professions; regulations requiring obligatory state checks on public and private facilities; regulations on the running and registration of societies and associations, particularly if they engage in practical activities; health and safety regulations, etc. In addition, apart from a small minority of natural dissenters, humans educated and brought up in societies in which extreme conditions, rather than political structures, impose a high degree of collective responsibility, see group co-operation as a necessary and virtuous part of their culture, and an essential part of their success in surviving. This has previously been witnessed in the founding of the Israeli kibbutz [16], in which a socialist style work ethic has been its defining characteristic. A more selfish type of individualism [17] is a luxury generally confined to environments where physical conditions do not regularly threaten instant death to members of the society. The effect of extreme conditions on this response to tyranny is to encourage acquiescence and thereby embolden authorities to increase the extent of their incursions. An extra-terrestrial society is much more likely to bow to centralised controls than a society on Earth.

The second response to tyranny is a Stoic withdrawal, but such a withdrawal from the affairs of society is in many senses illusory. Physical withdrawal, leading to possible mental withdrawal, is only possible on Earth because the commodities upon which basic survival depend, i.e., water, food and air, can be procured, in the right environment, by an individual. Where the conditions are appropriate it is possible to live a hermitic existence, completely detached from the apparatus of a state. In extra-terrestrial environments, no such extreme retreat is physically possible [18], since the basic commodities are all necessarily supplied by machinery, which is unlikely to be owned by an individual disconnected from society. The maintenance and spare parts necessary to keep such machinery running are beyond the scope of the industry of a single individual: they require connection into a network of people or, in fact, a society. With such a physically forced link to society comes an indivisible mental link.

But there is a much more fundamental problem with Stoic withdrawal. Even our individual thoughts are necessarily formed within the context of a society. A hermit in a forest might be convinced that the forest is beautiful, if they are contemporary hermits. But in the Middle Ages, at least in Europe, forests were regarded as forbidding, evil places, filled with wild animals and spirits, and not the domain of well-thinking people, which partly explains why there was so little compunction to cut much of the forest down. Our Middle Ages hermit might well have thought twice about locating his hut in the forest, and might not have thought much of the forest as a home. Very few of our attitudes and thoughts about things around us are developed independently. They are rather products of our particular age and culture.

The difficulty of achieving a social disconnection can be witnessed in overwintering populations in polar stations on Earth.

Those individuals who withdraw from the group, perhaps because of dissatisfaction with their social peers or their social situation, usually become unhappy. Ultimately, individuals in extra-terrestrial environments will be have to engage with their society and the authorities, whether a state or corporations. Inexorably, this will make them a product of their age and the wider social culture embodied in the polity. This can then lead either to one of the next two paths of resistance to tyranny, or to acquiescence.

The third response to tyranny is dissent. The problem with organised dissent in extra-terrestrial societies is that the dependency of individuals upon one another makes it difficult for dissent to occur without early detection. Furthermore, these societies will, simply by virtue of the period in the history of human civilisation at which they are established, exist when the monitoring of movements, conversations and social intentions will be relatively easy. The confined nature of human societies in the extremities of extra-terrestrial environments, and the need to live within pressurized habitats, make the deployment of this technology quite simple. The practical and technological limitations to dissent will be reinforced by the immense pressure to maintain the status quo against people who may be perceived, or who can easily be made to appear, to threaten social stability and thus the physical safety of other individuals. I am not arguing that democratic systems of open dissent cannot be created in extra-terrestrial environments, merely that the environment facilitates the degree to which those systems can be undermined and downtrodden by individuals or organisations that wish to do so.

The final response or buffer to tyrannical incursion is physical escape from the conditions of tyranny. In extra-terrestrial environments this is difficult to achieve. Regardless of how tyrannical the society may seem, the development of food processing systems, life support systems, oxygen production units, and all the other technologies of human survival is likely to make any conurbation of a reasonable size less practically demanding on individuals than trying to create a new settlement elsewhere on the planetary surface. In some sense, we find parallels between individuals moving from the highly developed societies of 17th and 18th century Europe to new colonies in the Americas. Like these early colonies, independent extra-terrestrial habitats will be more prone to failures than their originating settlements, and the founding individuals will find it problematic to achieve independence. In particular, individuals seeking to escape an existing extra-terrestrial settlement will not be separated from their source societies by, for instance, vast oceans, and they are thus likely to find it much more challenging to achieve physical separation, in materials and people, from the original settlements. Simply deciding to escape tyranny by moving to somewhere new on a planetary surface will not be easy, except for the most well connected and resource-rich groups and individuals, although it may not be impossible.

Individuals cannot simply run away, as civilians in an occupied nation on Earth might, either into forests or the desert, or some such relatively inaccessible environment. Survival 'on the run' in extra-terrestrial environments is not possible for extended durations of time given the lack of a breathable atmosphere, the lack of naturally available liquid water and the lack of indigenous plants and food to sustain oneself. Fleeing into an extra-terrestrial environment is like running into the Sahara desert, with the obvious exception that one will also need to carry oxygen to breathe. Such a severe restriction on movement is not unknown on the Earth: Australian convicts were kept in check by the convenience (to their captors) that if they escaped they were likely to die in the Australian desert, which allowed restrictions on movement to be implemented without requiring any such overt policy or practice.

Therefore no border stops or passport controls are necessary to impose restrictions on human movement from a settlement in an extra-terrestrial environment. The environment is a natural Berlin Wall. The fourth buffer against tyranny is at best weak and, for most of the time and for most people, it does not exist.

The conclusion we can draw, albeit a dark one, is that all four responses to despotism that are commonly exploited on Earth are either strengthened or weakened in favour of tyranny. As a result, society will express the potential for absolute tyranny.

Having discussed the effects of extra-terrestrial environments on buffers to tyranny, and explicated the reasons why people's options to respond to totalitarian behaviour are weakened, it is now necessary to explore how extra-terrestrial environments enable tyranny to be expanded [19]. The mere weakening of the buffers to despotism does not in itself imply that the march to tyranny would be any quicker or easier than on Earth, or even inevitable. However, I will now argue that extra-terrestrial environments not only weaken these buffers, they give succour to those people wielding power to expand their influence.

6. Extra-Terrestrial Liberty

For Berlin [20], 'negative' liberty meant the pursuit of individual liberty by removing those mechanisms that exert control over one's actions. Western liberal democracies pursue, for the most part, philosophies of negative liberty, by attempting to reduce the role of government in individual lives. A restricted sphere of negative liberty is created by tyrannies, in which encroachment into the lives of individuals reduces the number and scope of activities in which people consider themselves free, or at least able to make decisions that can be implemented independently of the state. Of course, by retreating into a core set of activities in which one is completely free, one is in the process of relinquishing liberty, as the scope of free actions is voluntarily reduced. This is in itself a form of slavery. Societies where the scope of negative liberty is reduced can be described as more enslaved, even if the people there may not describe themselves as such, because they have in fact escaped state slavery by retreating from those very activities in which control is exerted.

The crucial point is that the sphere within which negative liberty is possible is necessarily constrained by the environmental conditions under which one exists. The more extreme the environmental conditions, the fewer social activities can occur without collective oversight. More saliently, the people themselves may actually request such oversight, to protect their safety from others who would abuse it, with the resulting dangers. Some of these systems of monitoring can be found in societies on Earth. We cannot drive automobiles without safety checks. Our water must be passed through treatment works—life support systems if you will—that ensure that what we are drinking is safe. Indeed, even in some of the most mature terrestrial democracies, a remarkable quantity of basic consumables and resources come to us through systems of compliance overseen by the state. This is a form of control that most people accept because we consider it in our interest. We do not usually see such invasions of our liberty as tyranny, but rather as benevolent actions by the state to ensure our safekeeping. But they are incursions nevertheless, and while democracy is functioning such oversights need not necessarily concern us; or at least they do not worry most of the public, who are more concerned with having fresh water than more abstract thoughts about the allowable extent to which the state should have influence over their water quality.

In extra-terrestrial environments, spacesuits, water quality, food production, habitat pressurisation and so on and so forth will be subject to regulation by corporations or the state. As on Earth, perhaps many of these incursions will be regarded as acts of beneficence by the state in the interests of safety, and will be willingly accepted. But one fact is undeniable: the extent of negative liberty must be less in extra-terrestrial environments than on Earth, and quite significantly less. Even the air will be subject to quality controls and checks. Forms and permissions will be associated with the very act of breathing. No philosophy of advancing the domain of negative liberty, no clever sophistry, can change this truth, which is brought into being by basic survival needs.

An undeniable effect will be to expand the opportunities for tyranny. Where the mechanisms for central control are necessarily enlarged in their scope and diversity, a greater number of levers exist, and enable individuals and organisations to exert control and assume power. A reduction in negative liberty does not necessarily imply greater tyranny, but it certainly makes it possible. In extra-terrestrial environments, where centralised interventions must be frequent, how much weaker is freedom and how much easier is tyranny to enforce? We cannot know the answers until we undertake the experiment, but we can be fairly sure that the qualitative answer must be 'more easily'.

More insidiously, the restriction of the borders of negative liberty, caused by the apparent need to protect individuals from the irresponsible actions of others, can itself be perpetuated as a form of liberty. The use of alcohol in extra-terrestrial environments is one example. On Earth, the excessive use of alcohol may result in broken windows and arrests, but once the windows are repaired little damage has been done to society as a whole. Hence, although there is a negative social collective impact of excessive alcohol use, the prohibition of alcohol consumption of any kind is generally regarded as an infringement of civil liberties that the public will not tolerate. This is why, of course, attempts to do exactly this in the past have been met by black marketeering. But in extra-terrestrial environments, a broken window may imply depressurisation, and the instant death of many individuals. The potential impact on society of the irresponsible and thoughtless actions of individuals is greater, and it might seem justifiable to restrict greatly, or even prohibit, the civil liberty of alcohol use, in the interests of collective safety.

This principle can be applied to many diverse social interactions that could be construed as threatening people, and the prevention of which can be advanced as the protection of individual and social freedom through the process of restricting negative liberty.

Liberty encompasses the freedom that individuals have to actively pursue their own objectives ('positive liberty' *sensu* Berlin). An obvious mechanism by which this becomes practical is the creation of social mechanisms and institutions through which the 'active' pursuit of this 'positive' sense of liberty is made possible. Organisations established to act as conduits for the free expression of different points of view, or to act as means to achieve practical objectives, are not always liberal. Even in some of the most developed democracies, societies and organisations may become dominated by elite closed circles of people, and media channels can be influenced by moguls who use outlets to perpetuate specific corporate views. What prevents these incursions into the structures of liberty from descending into wholesale tyranny? In reality, very little. The subversion of democratic states, or states on the verge of democracy, into societies more reminiscent of dictatorships has many historical precedents. The principal mechanisms that allow individual freedom to triumph over the slide towards tyranny include the legally agreed freedoms that individuals have to establish competition against dominating organisations, and the culture that ensures that the freedom to create organisations is not then abused to destroy the very democratic organisations that guarantee that freedom.

In a society in which the freedom to organise and assemble institutions is protected by law, those organisations that distort and alter their environments, or the information they propagate, are likely to be usurped by institutions that reflect a different style of thinking, by the process of individual choice. However, these alternative visions can only be effective, and one can only assert them over the prevailing opinions with confidence, when one has sufficient information to be confident of their likely veracity. On Earth, to express many ideas and counter-opinions one does not need supreme confidence in the truth. If one's opinion turns out to be in error one gives up, accepts the viewpoint of the adversary and continues one's life. These opportunities to challenge, however, are central to the power of the individual to confront institutions.

But there is one social situation in which the individual's power is markedly reduced, even rendered completely ineffective against a collective body. Health and safety is one of the most effective levers of social influence and justified coercion, because it invokes the protection of people confronted with life and death situations. Consider, for example, an oxygen supply system on the Moon. The authority that runs such a system might seek control over a political dissenter by threatening to move him or her and their family to a new zone of habitation, on the grounds that the oxygen supply to their habitat is faulty. By doing this, they will remind these individuals who is in control of their survival, and coerce them through fear into mitigating their dissent, thereby creating a more malleable individual and reducing the challenge to collective authority.

Governing organisations have access to a vast realm of information that no single individual can hope to have[21]. They know, for example, about the oxygen demand, its rate of supply, the pipes that supply it, the maintenance history of the oxygen producing machines, and so on and so forth. For an individual to declare that the intention of the authority to move them to another habitat is for controlling political purposes, he or she must also have access to all such information, which they can then use to demonstrate that there is no safety concern. If they do not have access to this information, then it becomes a simple task for the authorities to portray them as dishonourable individuals inveighing against the hard work of other individuals who are working to secure their individual safety and the security of society [22]. They can be then be ostracised, and their general behaviour will be treated as disingenuous. However, to have access to all the information to convincingly uphold a complaint is never possible, because an individual can never know whether they are missing a single crucial fact that makes all the difference to their safety. Even armed with what they *perceive* to be all the information available, the individual is faced with a choice between allowing an incursion on their home and liberty, or taking the risk that their presumption of having full information is correct. Faced with such a choice, the individual is likely to opt for the former in the interests of caution, particularly in an environment where the other choice may imply death from a failed oxygen system. In the extreme case, this first course of action would be further reinforced in a particularly coercive, venal society where the individual might even be convinced of the capability of the authorities to engineer the failure of their oxygen system and their death, in order to crush dissent, even if their complaint was in fact justified.

The end point of this process, when applied across many activities in life, is a colony of automatons performing tasks for an extra-terrestrial authority, with their freedom reduced to a withered core of activities in the most private confines of their habitats. Extra-terrestrial environments make such an endpoint not merely a possible outcome, but a likely one.

This attack on liberty is made possible because the pursuit of individual safety can be made an unchallengeable requirement of a 'free' society [23]. Freedom from instantaneous death caused by the external environment is the common freedom on which all individuals should converge, and any social structure or plan that brings people closer to that reality must surely be praiseworthy? The removal of other freedoms to achieve the safety of society is excusable. From this position, the environment can itself become the instrument of positive liberty. In this way, and in a rather unique way, encroachment on freedom of thought and movement, in the interests of ensuring the protection of the freedom of the individual against the lethality of the environment, can be transformed into a justifiable and universalisable doctrine of control [24].

Unfortunately this approach receives succour from every major tradition of social philosophy that we know on Earth. From Grotius to J. S. Mill, the right to self-preservation has been considered the core of individual liberties [25], a point beyond which no state may go, and which every individual has the right to take it upon him or herself to secure—indeed, such a notion has even been referred to as a 'natural law' [26]. Even Hobbes' view of the necessity of sovereign control [27] turns on the right of each individual in a fight to preserve themselves. It is possible to spend much of one's life on Earth without undue concern for self-preservation. Apart from those unfortunate individuals who confront a burglar or gang, most people will not actually come face to face with the need to infringe others' rights to self-preservation. Fortunately, although the right to self-preservation is theoretically an unchallengeable right of all people, it remains, in a civil society, one sufficiently protected by the laws, and by regulations against various street crimes that might infringe self-preservation.

If, as has been traditional on the Earth, the right to self-preservation is also held to be a basic right of all people in extra-terrestrial environments, then the keys to despotism are handed over to those in control of society. Self-preservation is threatened on a day-to-day basis by the lethality of the environment. In such an environment, each individual does indeed represent a much greater threat to every other individual than on the Earth, because unpredictable and criminal actions against the infrastructure represent a continuously present and potentially catastrophic infringement on self-preservation. The authorities therefore have the excuse to implement draconian systems of control to protect the right of every individual to self-preservation. Worse than this, however, the people will voluntarily, in exercising their right to self-preservation, and to protect themselves, accept more far-reaching control over the lives of others [28]. Where death is a more likely outcome of criminal action, the Hobbesian state of nature, and the tendency to vigorously guard against it, becomes a more tangible reality [29].

7. The Influence of Population Size

The size of an extra-terrestrial society will have an influence on the characteristics of liberty that have been discussed, although these influences are cosmetic in the broader scheme of the principles of liberty I have discussed. Whatever the size of society, and no matter what the scale of infrastructure, the outside environment is fatal. The boundaries of social control are set by that environment but, within those boundaries, the manner in which it is actually implemented will vary from society to society in subtle ways.

Small societies nurture a greater degree of camaraderie, which will partly offset the sense of desolation in the extra-terrestrial environment [30]. Inter-dependence between people in small groups where people know and communicate with each other more regularly than in larger societies makes authority less liable to abuse a population since its mechanisms are more visible. On the other hand, it is more difficult to remain anonymous within smaller groups of people, and the mechanisms outlined in this essay, through which people will exert control, can be more effective. The fewer people there are, the less is the diversity of opinion and the harder it is to find like-minded individuals who share a common concern about a negative aspect of society. The lack of anonymity within a small social group, added to this lesser diversity of opinion, increases the intellectual courage that is required to dissent from the agreed norms and customs, and thereby reduces the number of people who are willing or even able to dissent; negative liberty is more likely to be restricted. However, conversely, positive liberty may in some situations be more difficult to transform into a tool of repression. A small group of people has less inertia, and, in very small groupings, even individual opinions may be incorporated into social norms in order to accommodate as many people's wishes as possible. In such situations, a difference of individual opinion may more easily influence the overall vision of the direction of the society.

The case of small groups of people is most likely to pertain to extra-terrestrial environments. Not until large economic resources and supply chains are mastered are extra-terrestrial societies likely to exceed many thousands of people. We should consider that, at least in the initial stages of social development, the development of society will follow those expected of small groups of individuals.

Larger societies generate a greater degree of social fluidity, within which it is easier for individuals to avoid central oversight, and easier to physically escape from groups or individuals that one might wish to avoid [31]. The larger a society, the easier it is for individuals to take it upon themselves to increase their own sphere of negative liberty. The larger and greater number of habitats and spaces required to support a larger society necessarily allow for a greater redundancy in the equipment to support survival, and the need for central control is diminished. More developed corporations and economic systems themselves create redundancy in the potential supply of critical components for life support equipment.

Despite these possible effects of increasing population size on the power of corporations and central production, it is counter-intuitive, but historically the case, that sometimes the larger the population size, the more easily totalitarian control can be imposed, because the mechanisms of control become more obscure, and the less information any one individual can obtain. The mechanisms of centralised bureaucracy, like tentacles, permeate organisations and institutions across a wide area, and it becomes more complex for an individual or even a group of individuals to influence this machinery or attempt to alter it. The complexity of such a task is also made more intimidating by the sheer inertia of organisations that command obedience and resources, across both the physically horizontal and administratively vertical structures of society. This explains why very large geographic expanses on Earth have been successfully ruled by tyrannies. As extra-terrestrial societies grow, so the anonymity of the individual will increase, but the opportunities for the expansion and consolidation of networks of administration will also multiply.

The historical trajectory of extra-terrestrial settlement will itself act as encouragement to the development of a despotic character to larger settlements. The necessity for central means of distribution and production of vital commodities in the early stages of settlement will increase the chances that this bureaucratic 'baggage' of collectivist philosophy will be inherited by all subsequent social structures as they develop and expand. However, as the development of early European colonies in the Americas into the United States testifies, the formalisation of rights, constitutions and other political apparatus can be used as a tool to overcome the early necessity for heavily centrally planned societies.

8. The Bearing of the Political and Economic System

There is a widespread habit, even in the some of the finest social philosophy literature, to make a distinction between democracy and more autocratic forms of government, whether theocratic or politically based.

Now it is clear to even the casual observer that the vast majority of important business, even in a democracy, is carried out by unelected forms of central control. These vital functions, exercised by military institutions, intelligence agencies and ministries for environment, education, employment and so forth, are organised by hierarchies of institutional structures that are not subject to the oversight of mass popular will. I will call all of these layers the 'oligarchic core' [32].

I use the word 'oligarchic' loosely to mean institutions and people not subject to external democratic censure. Many institutions do employ internal democratic processes to elect new officers and leaders. Conversely, it is also true that some institutions may resemble autocracies more than oligarchies, and that certain corporations may even exhibit elements of plutocracy. Nevertheless, the word 'oligarchic' usefully encompasses the general notion of the relatively small body of people who run each institution that carries out vital state functions. I should also emphasise that, by using the word 'oligarchic', I am not implying that these institutions are necessarily actively anti-democratic, or engaged in conspiratorial plans to undermine democracy (although in historical instances in which democratic states have been overthrown by military juntas, or similar forms of governance, the individuals responsible for these actions have often hailed from what I refer to as the oligarchic core).

The oligarchic core is generally hidden because it resides under the veneer of the democratically elected leaders and ministers who provide the defining political characteristics of a democratic society. The best analogy I can conceive of is an orange. Its outer peel (democratic systems) is visible to the populace and the wider world, but it covers the interior structure (oligarchy), where all the essential functions lie. The development of seeds for the following generation (education), the segmentation and compartmentalisation of functions (government institutions and ministries), the movement of fluid (movement of people and information between these segments) and the growth of the supporting material (the political institutions of the economy) are fundamentally similar, whether the peel is on or off the orange, but the peel is the visible part of the orange.

Before the reader should infer that I am somehow cynically implying that democracies are only marginally better than oligarchic and possibly totalitarian states, I should point out that we should be pleased with this fact about democracies. For if all our leaders of institutions were swept away every four or five years, and replaced by new officials, we would have a very unstable, and probably dangerous, society. It is a necessary characteristic of a stable society that many vital social functions should be organised and supervised by people with long periods of experience that make possible rational evolutionary development and progress, rather than replacing people every few years, which would result in short-term, regular revolutionary and unpredictable transitions.

But I think it remains, nonetheless, the most striking paradox of democratic systems that the stability of democracy rests on the unwavering stability of these hidden layers of oligarchy. It is also true that the idea that there is in fact a categorical difference between a democracy and other forms of undemocratic central governance is almost certainly delusory.

This analysis does bear on the question of economic organisation. A market economy will necessarily generate different types of political institutions, such as trade associations, than a centrally planned economy, and these same institutions will affect the nature and type of political environment that results. There seems to be no obvious reason why both market and centrally planned systems of economy could not be created in the space environment (although we can observe that the vast spatial scales of outer space make it likely that many extra-terrestrial economies will be autarkies) [33].

Market economics has an influence on the oligarchic core, because its fluid wheeler-dealing and the lack of state-directed centrally planned economic goals disconnects a great deal of economic activity from the oligarchic core and makes much of it a separate layer, influenced only by the overarching legislation that controls the conduct of corporations [34]. In that sense, one lever of power of the oligarchic core is dramatically reduced in scope and makes tyranny less possible, less easy to attain. We might say that the market economy provides succour and health to the democratic veneer, both by encouraging a greater scope of individual free thinking, given the ability to form corporations, and by reducing the opportunities for governing institutions to exert economic power. However, it is clear that corporations themselves could replace this power, thereby merely transferring economic control from the oligarchic core to the oligarchy of large corporations [35]. The extent to which this latter assertion is true, and the relationship between the institutions of governance and social management (the oligarchic core) and corporations, are matters for discussion elsewhere. For my purposes in this essay, it is sufficient to understand the existence of an oligarchic core in the political institutions of state and society, regardless of whether the economic environment is dominated by central or market models of planning and development.

The link between democracy and its hidden oligarchy is a two-way flow of information. No democratically elected leader can have the full breadth of information necessary to run competently the affairs of state, from internal domestic matters ranging from education to the environment, as well as foreign affairs that on Earth encompass the full panoply of interactions with many tens of nations at any one time. One function of the hidden oligarchy is to provide advice, and this guidance, proffered by people who have remained in their positions over many decades, is valuable. This is particularly so in dealings with foreign powers, where the history of a nation and its culture may intimately affect its view of other nations and its behaviour, and a poor judgement by an elected official, more especially a very senior one, may cause an international crisis, even war. The hidden oligarchy provides knowledge and continuity from government to government, and this reduces the chances of such errors in many of the activities of the state, both foreign and domestic. It is, if you will, a repository of long-term political and economic experience upon which the democratically elected officials must draw to guide them in the running of a society. It is also the machinery by which the banal but vital day-to-day functions of government must be performed, and this machinery too requires many decades of practice to perfect.

In reverse, the function of the democratically elected officials is to provide fresh policy, and to oversee the oligarchy. Their influence is two-fold. Firstly, by the very fact of their positions being underpinned by a widespread popular mandate, they ensure that the democratically elected veneer of society has authority over the oligarchy, morally as well as politically. This has a subduing effect on the worst excesses of the oligarchy, both by keeping it subordinate to a higher authority under the regular judgement of the people, and by exerting a psychological and practical influence over the leaders of the oligarchy, curbing the extent to which they are likely to pursue autocratic social programmes that do not command popular support, because these programmes would result in the removal of the democratically elected leaders by the people. Second, their fresh vision and desire for new policy directions may compel them to change leaders within the oligarchy, thus preventing stagnation, and denying that thirst for the status quo and the retention of power that is the inevitable course of events with leaders within any autocratic institution who have their power and influence to maintain.

So it is not true that democracy is merely some sort of facade, a window dressing for unelected institutions and leaders. Democracy has a vital role to perform in controlling the worst tendencies of the oligarchy. In my orange analogy, it prevents the interior from drying out, and maintains the free flow of fluid or information between the segments of the state. It is however, a thin veneer, at least in terms of the numbers of personnel directly subject to the process of popular democratic scrutiny. It would not be far wide of the mark to say that there is no categorical difference between democracy and oligarchy, that democracy is a sub-species of oligarchy.

I have taken this wide diversion because it is necessary to understand that, whatever the structure of governance in an extra-terrestrial environment, at its core it will be oligarchic, as it is on Earth.

This analysis is important because some may claim that the validity of everything I have said in this essay rests on the type of government that is formed in an extra-terrestrial environment. If it is democratic, then surely all concerns about centralised control and the retraction of liberty are void, since the people will simply object, cast out their leaders and replace them with people who push back the boundaries of central control? Now as we have witnessed during the ebb and flow of socialism in western democracies since the 1960s, there is no question that there is plenty of room for protean economic and political approaches under the democratic veneer. If it also transpires that democracies are established in space, then the people will have the opportunity, through their elected leaders, to choose to insist on the repeal of autocratic policies to the maximum extent possible.

But if it is the case that, because of the natural environment, the retraction of negative liberty and the use of positive liberty as an instrument of control is more likely to be adopted than on the Earth, regardless of the type of government established, then it is also true that no amount of democracy will change the likely strength of the oligarchic core in extra-terrestrial environments. To return to the analogy of the orange, an extra-terrestrial society can be viewed as a dried orange. It maintains a democratically elected body of leaders, but the lethal and physically restrictive effects of the environment, which require a greater oversight of social affairs, accompanied by a more subdued population, are equivalent to a hard and dried core at the centre of the orange. The seeds still develop, the segments are still clearly defined, but the centre is less fluid, and the flows of information and people made possible by a free and open environment are restricted. In the two-way flow of information between the democratic institutions and the oligarchy, the formation of a rigid and less fluid oligarchy inevitably exerts a more powerful influence over democratic institutions and their leaders. Their policies become more difficult to implement, and subject to more due consideration before they can be practically implemented, and the leaders themselves become more conservative. In the confined societies of outer space, poor judgements by democratic leaders may more easily result in social disaster or death, making them more cautious, but also strengthening the power of the leaders of the oligarchy, by confirming that their experience endows them with better judgement. As with the orange, the hardening of the oligarchic core inevitably causes the outside peel of democracy to wither and retract.

The rigid, unfaltering nature of the core is facilitated by one of tyranny's most loyal friends since the birth of civilization— petty officialdom. It is not that functionaries conceive bad laws and social programmes, it is that they enable and bring to fruition the plans and visions of those who do have the imagination to conceive of them, of evil that would otherwise find no avenue for its expression or spread throughout social systems [36]. Petty officialdom finds its way to high influence in societies where certain social requirements are vital for the functioning of society, but require nothing but blind obedience to a higher authority to implement. In any environment where there are a substantial number of actions that require this official oversight, then officialdom will find an open field. The number of these areas of official involvement in extra-terrestrial societies, as I have already noted, will be much greater than on Earth, particularly in matters that deal with survival and health and safety. Amongst these, we can count the need to inspect habitats for potentially fatal depressurisation leaks; the need to carry out vigorous fire safety checks in such confined environments; the need to implement strict health checks, both in private habitats and in the storage, sale and processing of food, to prevent disease outbreaks that will spread rapidly in enclosed settings; the requirement for general checks on all manner of survival items including food production units, tubes carrying air to breathe and electrical systems that regulate the basic supply of air and water, *ad infinitum*. Both the self-motivated desire of officials to arrogate these socially beneficial checks to expand their power, and the opportunities these requirements offer central authorities to encourage officials to build their own empires and inspissate bureaucracy, in exchange for obedience to the central authority, will make petty officialdom one of the most powerful instruments for solidifying the oligarchic core of an extra-terrestrial society.

Therefore, whatever the nature of governance in extra-terrestrial environments, the continuity of both experience and knowledge manifest within the oligarchic structures of governance will exert a greater influence over liberty than on the Earth, in the direction of tyranny, regardless of whether the leaders of the society as a whole are democratically elected.

9. The Character of Extra-Terrestrial Freedom

We have established that extra-terrestrial environments both restrict the efficacy of the buffers to tyranny, and provide the conditions in which dictatorship can take advantage of those weakened buffers. We now return to the question: what will be the extra-terrestrial concept of freedom?

Freedom can be perceived by people in two distinct ways, which I will call here 'comparative freedom' and 'real freedom'. Comparative freedom is a freedom felt when another group of people is perceived to lack that freedom, or vice versa: the unnecessary repression perceived when another group of people seem to be freer. Comparative freedom is often used by politicians to encourage people to feel fortunate that they have the freedoms that the state guarantees for them. Comparative freedoms may be valid from some perspective, particularly if a people has been rescued or saved from a type of social system that has taken freedom away from other people [37]. However, it a dangerous type of claim to freedom, because it allows a tyranny to remove liberties to the maximum extent possible, whilst preserving some freedoms not possessed by other peoples, in order to uphold the illusion that liberty has been maximised. And, of course, comparison with other societies provides no insight into the potential freedom one might have, since other societies may simply be tyrannical in the extreme and one's own society not much better; but, even so, sufficiently better to enable those governing it to legitimately claim that it is more free, even though it is still far from the freedom it might achieve.

What I call 'real freedom' is the freedom people think they have independent of any comparison: their internal sense of what activities in their lives they think are not under excessive scrutiny or control by others. To a certain degree this is subject to the social conditioning of one's expectations, and it therefore has a comparative element. In a society where family or community bonds are very strong, control over the manner in which one's earnings are distributed may be tolerated to a greater extent than in a society where family obligations are less great, for example. However, there is a distinction between freedoms one identifies as absolute in one's life, and freedoms that derive from an explicit comparison with other people.

How will freedom be viewed by people in extra-terrestrial environments? It is not possible to speculate on the specific types of social customs and norms that might emerge in a new society, but it is possible to establish some likely trajectories of the basic concept of freedom.

There are certain physical factors in the extra-terrestrial environment, no matter where within it one lives, that will give people cause to believe that they have greater comparative freedoms that result from the physical environment than people on the Earth. Interplanetary space, the Moon and Mars, for instance, lack devastating earthquakes. Although not everyone on the Earth experiences earthquakes, people in space will no doubt perceive these natural disasters as a terrifying threat: a threat that is unpredictable, and likely at any time to overwhelm and destroy lives [38]. Floods and tsunamis will be viewed in the same way, as fickle and dangerous anomalies of nature that belong to a planet with oceans and copious quantities of liquid water. Although Mars has planet-encircling dust storms, the atmospheric pressure on Mars ensures that these storms will not inflict the same destructive influence on property and people as hurricanes or tornadoes do on the Earth. Without biospheres of any great scope, few extra-terrestrial environments will threaten people with viruses, wild animals or other biological agents, except those intentionally or inadvertently brought from the Earth. In all these respects, people in space will feel comparatively free.

But in many other respects they will lack physical comparative freedoms in a most fundamental sense. As pointed out earlier, the most basic and incontrovertible of these is the freedom to wander outside and breathe the air. People will not find the variety of colours and wildlife found on Earth, and to them this may suggest that they live in a repressive society.

These physical comparative freedoms do not necessarily engender dislike for one's government, and it is important to understand the distinction between freedoms one simply cannot have and freedoms one can. When there is a drought, and water is limited, I do not hate my government for the fact that it has not rained. That would be irrational [39]. But differences in the physical environment can bring policy and practical limitations to freedom where the boundaries are ambiguous. I may not dislike my government for restricting my use of water during a drought, because I consider this a responsible action to take, and I am inclined to obey the government in the interests of preserving water supplies. But I may well begin to resent my government if I believe that the restriction is too great, or if it continues too long, even though the government may actually be correct in its determination of how long to keep the restriction in place. And this is the point: physical differences in the environment, manifested as differences in social and political comparative freedoms, can give rise to a sense that others are unnecessarily less free, or unjustly freer, when the rules and laws that result from those environmental conditions do not meet with the agreement of people. It is likely that the fewer physical freedoms that extra-terrestrial environments allow people will lead inevitably to a sense of less comparative social and political freedom, regardless of how tyrannical or free a political system in an extra-terrestrial environment tries to be. As a result, the extent of disagreement over the necessity of the resulting regulations will be greater.

Given the likelihood of this condition, we can expect any apparent social superiority in extra-terrestrial environments to be an attractive focus of comparison, both for people themselves, to justify pride in their society or to provide legitimacy to their lives in that society, whether they are there by choice or not; and for the authorities, who will, against the repressive extra-terrestrial environment, seek any measure of beneficial comparison. Extra-terrestrial societies, will, at least for a certain duration, be free of political threats. Nuclear terrorism, mass conflict, famine and other horrors associated with a densely and highly populated planet, or the thousands of years of cultural disagreements on Earth, will be absent in space, although we cannot be sure that such things will not emerge as the societies grow in size and develop their own cultures [40]. Some of these terrestrial problems and political crises may make any type of tyranny in space appear mild in comparison to what is on offer on the Earth.

The real freedoms that people hope will be defended by formally agreed rights will be defined and influenced by the societal response to the extra-terrestrial condition. On Earth, societies define freedoms, many of which we might consider to be 'real' freedoms under my definition, and they may establish a set of rights to protect that freedom. In America, for instance, the Bill of Rights [41] sets out certain rights, inalienable rights. All of these rights sit within expectations, in fact assumptions, about the grander freedoms to which people are entitled. Nowhere within the American Bill of Rights is the freedom to breathe air listed. Why is this not the case? The reason is that no matter how bad despotism on Earth may become, a government cannot take away the air. On the moon or Mars it could do exactly this. Nowhere in the American Bill of Rights is the freedom to drink water enshrined. No matter how pernicious the government might become, people could at least collect water from rain without state interference. No such freedom exists on the moon and Mars.

It may be possible, and it may be required, for extra-terrestrial constitutions, whatever form they take, to protect certain rights: that all people have the right to access oxygen to breathe; that all people have the right to access drinkable water from water acquisition infrastructure. In pursuing the rights to breathe and drink water, will people forget to defend less fundamental rights? Or is it not likely that, in return for protecting these rights, the government may waive or eliminate other less obviously fundamental rights, which on Earth would be considered fundamental? It is not necessarily the case that the more basic the rights that are required, the more other rights are downtrodden, and the smaller the diversity of rights that is protected. But there can be little hesitation in saying that in societies where people must constantly work to uphold those rights that protect the basic freedoms associated with survival in an inherently lethal environment, other freedoms and rights may seem trivial, or superfluous. More to the point, they may be treated as such by those who run society.

10. Children and Extra-Terrestrial Freedom

It is in the education of children that we may expect the greatest repressive influence on development. Wayward children are a problem to society in many ways, including the potential damage that they might do to vital equipment such as life support systems, and other private and public property. There is not only the need to educate them to be civilised and responsible, but their parents' will fear being held responsible for rearing ill-educated offspring. For all the reasons I have adumbrated in this essay, these fears would be much greater in extra-terrestrial enclosed environments, even though we also find them on the Earth. Society will demand subdued children who are serious-minded and predictable, in preference to wild, self-confident and independent-minded youth, since this predictability will secure safety [42]. Reinforcing this subdued character will be the experiences through which youth will be nurtured and developed. These children will not play in fields, or run freely along streets with friends or enjoy the expressive freedom that inheres in the environmental conditions of a planet where there are few limits to outside movement. Their playing will be done in pressurised habitats, and their fields will be, at best, growth units with artificial landscapes. Their friends will live in accommodation and surroundings much like theirs. They will be born into an environment of sensory deprivation. It would not be inaccurate to say simply that children born in space will be the first humans to be reared in cages.

Society has an incentive not to expand their vision of freedom to the extent found on the Earth, since this will breed discontent or, if they have the means, mass emigration to the Earth. The wide open expanses of the Earth, the breezy blue skies and the ease with which, even in large cities, people may seek the stimulation of new surroundings, are more likely to become the best guarded secret of extra-terrestrial societies. Worse, they may be portrayed as insignificant and valueless facts about the Earth, and those who feel enticed by this vision will be persuaded strongly of their error. The extent to which this will succeed is an open question. Perhaps, like the attempts of Soviet leaders to convince their populations of the inherent evil of western freedoms, it will be impossible to hide or deny the attraction of what is distant, enticing and novel. Indeed, the inaccessibility of the reality of Earth to children born in extra-terrestrial environments may even encourage them to believe in some sort of terrestrial utopia. Like the beauty that blind people imagine in the Earth's environment, which often renders them disappointed if they regain sight, the idea of blue skies, rivers and oceans will be pictured as a paradise by those who have never witnessed it.

Of course, irrespective of the extent to which the Earth is a paradise, the extra-terrestrial environment, whatever its limitations, will still be home, and these individuals will never be able to deny their ties to their birthplace, or its call. In that, they will be loyal to their home.

However, there is an equally likely outcome. Unlike Soviet restrictions, the restrictions on their freedom of thought and the development of their minds will not be wholly political, but rather imposed upon them by their environment, with no human hand to blame. Therefore, we might expect the children born in these environments to have a genuinely restricted notion of the boundaries of liberty. They will be less free, but they will expect less freedom; they will be the children of a most fundamental natural type of tyranny. I call a situation where individuals are restricted by their natural environment and the safety checks and culture that result from this a 'natural tyranny'. A subtype of natural tyranny is that in which individuals are not aware of the limitations imposed upon their thought by the society that results from their natural environment, since they consider it normal: they have been brought up within these restrictions. This type of tyranny might logically be called 'cryptic natural tyranny': repression resulting from the political and social environment that is imposed as a result of the natural environment and hidden by its apparent normality. This cryptic condition will only be revealed when an individual travels to a more open natural environment, *and successfully assimilates,* thus having the natural tyranny of their previous condition revealed to them.

Young people born into natural tyranny may even learn to resent, and to feel threatened by, the decadence of the Earth and the culture its free environments spawn. They may, quite simply, not understand it or appreciate it. Like the introduction of a bushman to the streets of London (if such a thing has happened), Earth may confront them with experiences simply beyond their comprehension. They may, in some ways, understand the freedom it offers, but they might equally hate it, and yearn for the safety of habitats and the predictable hum of life support systems. They may even feel some sense of pity for the people of the Earth in their vast, unpredictable, often dangerous and uncontrollable world, whose interconnections and deeply dependent links with the biosphere make their futures beyond the plans and predictions of even the best engineers and scientists: a vast spaceship whose workings are the diverse and fickle mechanisms of the blurred mysteries of over three billion years of biological and geological evolution.

From birth, the extra-terrestrial society will restrict the boundaries of liberty, both through the natural environment, and through the political behaviour that it causes. People may be aware of this restricted liberty, but it is just as likely to be so normal in the upbringing of people in space that they will be naturally less free.

11. Religion and Extra-Terrestrial Freedom

It is impossible to engage in reliable deliberations on the influence that religion may or may not have on extra-terrestrial freedom. Its manifestations are infinite and varied, and in any society it will affect both individuals and the population as a whole in ways that depend upon the degree to which religious belief permeates everyday lives. Although in western society today religious belief is widespread, in many nations it has little pervasive influence on the day-to-day affairs of state decisions, and even on many individual lives. We cannot meaningfully predict what sort of religious beliefs might emerge in extra-terrestrial societies, and what subtleties might inhere in them. Nevertheless, it is worth making some observations on this aspect of human society and its relation to liberty.

Here I am chiefly concerned with religion as it reflects worship of a God or gods. The acceptance of a religious point of view, such as Christianity, primarily as a means to provide a moral foundation for one's life as it relates to a set of perceived values, such as honesty, selflessness, etc. rather than the worship of a deity *per se*, is mainly a private matter. It would seem to be the case that, unless those moral values come into direct conflict with the values laid down in the society, this sort of religious conviction need not attract any particularly strong feeling from a society. The caveat to this observation is of course the level to which an individual who uses religion in this way actively attempts to convince others of the veracity of their values, and the extent to which they try to proselytise, or to extend the link between their values and the specific doctrine from which they believe these values emanate.

Religious belief may provide an expression of freedom for individuals when those individuals are able, and are allowed, to accept these beliefs of their own volition. Religion becomes a tool of tyranny when the state or religious authorities require a population to advocate and practice, or even indeed merely to agree with, a specific religious point of view; or when religious institutions themselves, either directly or aided by a state, impose substantial moral authority over large numbers of the population. Unfortunately, even on the 21st Century Earth, this second type of religious influence is still widespread in many societies. A central challenge facing the extra-terrestrial state will be to manage religion in such a way as to encourage its propagation in the first manner - through individual choice.

In the perpetual struggle to make light of the purpose of life, and to provide relevance to one's existence, belief in a higher authority or god that provides a deterministic undercurrent to life has been a pervasive feature of human society. All religions are in some way motivated and ultimately made viable by this deeper need that many humans have. In extra-terrestrial environments, the physical and political extremities that emerge will create conditions that are ripe for religion to flourish. There are multiple complexities to the emergence of a specific religious belief, but from a more general perspective there are two factors that will encourage religion of any kind in outer space.

First, it has been observed that, in extreme social and environmental conditions, people gain comfort or solace through religion. Against the constant life and death struggle of survival in extra-terrestrial environments, religion will find no more welcoming home. It will provide a distraction from the tedium and lethality of the environment, and be a source of hope and encouragement that, whatever the raw reality of the outside extremities, there is a spiritual retreat into something more beautiful and personal inside, which may itself link to an omnipotent presence that has no universal boundaries. What better place to convince people of a heavenly paradise than a dusty, barren and deadly radiation-baked wasteland, or the utterly empty, black, sterile and lethal infiniteness of interstellar space?

Second, the emergence of religious sects and beliefs will be made easier by the paucity of prior social structure or precedence in extra-terrestrial environments. There will be very little previous social order (except that which they bring with them) to set boundaries to what is religiously acceptable: a new society or group established on another planetary surface, or in space, will be practically a socially blank canvas. Although guided by the fundamental constraints of basic human behaviour, it will be open to entirely new concepts and ideas in many social areas, including religion. For religious individuals who see an opportunity to start a new creed or group, isolated extra-terrestrial societies seeking a meaningful vision of their purpose beyond their economic or political function will be not merely an obvious outlet. These societies will probably enthusiastically entertain any externally suggested social diversion or novelty.

Although religious belief more often than not constrains human thought, for some people it will be regarded as liberation from the extremities of extra-terrestrial conditions, just as many on Earth regard their religious devotion as an escape from daily troubles and pressures [43]. As the boundaries of liberty retract in extremes, so religion provides, even if in practice only in an illusory way, an expansion of the sphere of liberty, and a direct and powerful way in which an individual, completely independently of others, can feel a sense of a command over their destiny, and the opportunity to engage, on their own terms, with an authority higher even than the extra-terrestrial state.

Despite these open possibilities, one should note that the relationship between the state and religious belief is likely to play an important role in the character of the religious influence on people. As in the 20th Century communist states of the Earth, a greater degree of state control, made possible by the extra-terrestrial environment, may also be an instrument that restrains religious institutions. For all the reasons discussed, organising and coordinating religious worship will be difficult without it becoming visible. If strong religious groups emerge after the formation of a state structure, they may be considered more a threat to stability than a tolerable diversion for the population. In this case, it is not necessarily true that religion would be permitted free expression; in many social structures it may be more constrained than it would be on the Earth, and might become confined to a personal activity undertaken by individuals, possibly even in secret. However, if the formation of the consolidated state structure follows the emergence of a strong religious belief, then we could even imagine theocratic extra-terrestrial states, where religious belief, as a social mechanism of order and peace against the extreme environmental conditions, is intimately embedded in the function of the state, and is used as an instrument of liberty.

The reality of the existence of a god is something for individuals to decide, but it does seem likely that, whatever one's point of view on this matter, we cannot deny that religious beliefs of various kinds are likely to emerge in an extra-terrestrial society. They will affect the view of freedom, if not at the level of the state, then certainly at the level of the individual.

12. Private Property

At least since Locke's persuasive argument concerning the power of the individual to transform land into private property by toil [44], the right to private property and its distribution to others, according to the wishes of its owner, have been considered a fundamental component of a liberal or libertarian society. This central tenet is protected with a type of zeal: the violation of the right to private property is one of the worst abuses that can be committed against people within the philosophical milieu of modern liberal democracies.

The unequal distribution of property among people can be accommodated within a society where poverty rarely results in death. Those who have cannot be regarded as a threat to society simply because their ownership of unequally available resources implies that others have not. But can poverty be tolerated in extra-terrestrial environments where there is no place for the impecunious to go, apart from pressurised walkways, habitats and other public areas, which, by the nature of the extra-terrestrial environment, are confined spaces? Can poverty be tolerated when the criminals it might rear could threaten unimaginable devastation to enclosed societies in lethal environments?

A large extra-terrestrial society, where resources are plentiful, may be expected to operate much like a society on Earth, in terms of the availability and distribution of resources. However, unless those societies have population sizes on the scale of nations, the want of some resources and property by some may represent both a potentially fatal lack for them and, by extension, a danger to the rest of society. A lack of these resources relating to air, water, food and other basic commodities will encourage people to levels of desperation far greater than could occur on Earth, because the 'natural' environment does not provide a retreat to which the poorest people can escape: you cannot sleep in a cardboard box on the foothills of Olympus Mons. An extra-terrestrial society cannot afford to have outcasts from society who lack the property essential for their survival, nor can it afford an underclass of criminals bent on achieving vengeance against a society that they perceive to have denied them certain vital property.

On Earth, the welfare state need not necessarily imply gross interference with individual freedoms, and in extra-terrestrial societies it is plausible that a type of welfare state, although much more extensive than those on Earth, could protect the least successful members of society from the dangers they pose to themselves and the rest of society, and at the same time protect the rights of individuals to private property. But it is a necessary condition of extra-terrestrial environments that, in order to control crime, reduce poverty and ensure that the basic survival needs of all individuals are met, the distribution of private property is likely to be monitored and controlled within a far greater diversity of activities and systems of production than they are on the Earth.

The provision of private property may be accomplished by many mechanisms. As on the Earth, vital supplies can be provided centrally, in which case there would be a centrally planned government or corporate distribution, and the implication for the discussion here is that private property would be extensively regulated. The need to achieve parity between supply and demand in an emerging extra-terrestrial society will encourage the development of a type of command economy, although the economy could well be organised and directed by democratically elected officials. Alternatively, private property could be supplied by private enterprise and corporations, with supply and demand regulated by simple market demands. Even during the earliest stages of the construction of extra-terrestrial societies, private enterprise could efficiently supply many types of property.

Private property is a realistic objective of the expression of freedom in an extra-terrestrial society, although initially it will be hard to achieve in small extra-terrestrial societies, for all the commodities that people might seek to acquire as private property. The need to accomplish a fair and basic distribution of vital resources amongst all members of extra-terrestrial societies will be an inducement to political control, because the requisitioning of property to support the vital needs of others can be justified by the requirement to prevent danger to society: a system that is more difficult to defend on Earth.

It is in the acquisition of private land, the object of Locke's original attentions, that the nature of freedom is more obscure. Some thinkers, who have given consideration to space as a frontier, have suggested that other planetary bodies will be like the early 18th and 19th Century frontiers on Earth, or the expansion of early American society into the West; but without, however, the obvious ethical complications that attended the displacement of indigenous people in the course of colonial conquests. Individuals and corporations may lay claim to extra-terrestrial land as a commodity, as they do on Earth. They will do this within whatever regulations and rules for such acquisitions will eventually emerge from the Outer Space Treaty, the Committee on Space Research's (COSPAR) planetary protection guidelines [45], and other policies and laws yet to be developed, as the human settlement of space unfolds. However, another perspective that derives from my previous analysis should be considered. Many extra-terrestrial environments will take on an image not dissimilar to medieval views of forests, as hostile places quite inimical to freedom. Except for the acquisition of land for commercial use, either by individuals or corporations, most individuals may regard land as a source of despotism and repression, and not, as on Earth, a source of liberation. In that sense, people may even develop a type of anti-Lockean philosophy: land, transformed by collective effort into habitats, life support systems, greenhouses and other facilities that support life, will be regarded as the fertile location of comparative freedom, but land owned by the individual, unless transformed by this collective effort, will rather be symbolic of the overwhelming and quite impractical toil that a single individual must undertake to secure independent freedom through the possession of land. Locke viewed the difficulty of transforming barren land into productive land as the very source of an individual's claim to that land, as a liberating mechanism and a path to the private acquisition of property. It may be a powerful irony that, in extra-terrestrial environments, the sheer scope of toil needed by an individual to alter land into viable private property may transform the Lockean vision of land as a source of individual liberty, into one of the most potent and convincing philosophies of the power and necessity of collective cooperation in the attainment of liberty.

13. The Tendency to Tyrannise

I now briefly address an assumption that has underpinned this essay. It is the view that when the supporting culture and social foundations of liberty are weakened, society will have a greater probability of sliding towards despotism. In other words, dictatorial conduct will tend to expand to fit the maximum moral, legal and practical space that it can occupy unless very assertive legal and moral attempts are made to halt it [46].

Many people would regard this as cynical, and a sorry view of the beneficence of the human character. However, I view this assumption to be correct given the basic human urges.

All people, at some stage in their lives, recognise their mortality. As a result, many realise the need to achieve, in the course of their lives, something that attracts some sort of recognition. From this derives the sense that their life has been of value. This is not a ubiquitous characteristic. Many people are content to live their lives without any specific ambitions. But such people are not generally influential in the direction that society takes towards liberty or tyranny, although of course they may well represent the mass of people enjoined by freedom-seekers or tyrants to join their cause. For the remainder of humans, there are two methods to achieve recognition in society. The first is to contribute creative ideas and projects that change society in ways that are recognised simply by their impact on human lives, or on the state of thinking in any particular age. In this category we include Kant, Newton, Picasso, Keynes, and other philosophers, scientists, artists, economists and so on, but of course it is not necessary to be as historically established as these individuals to sit within this category.

The second method of achieving recognition is by accumulating power over other humans, and thus increasing one's influence on the affairs of a greater number of people. This influence provides a sense of significance in the course of events, particularly when the acquired power and influence obviously impinge on the behaviour of other human beings. For the vast majority of the population, it is this second means to recognition that is practically possible, for the simple reason that to belong to the first category requires unusual intellectual skills, which not all possess, and which cannot easily be learned. The reader may properly point out that the two groups of people are not mutually exclusive. Many scientists and artists, for instance are, and have been, notoriously egotistical, seeking fame and influence for its own sake. It is probably not unreasonable to contend, however, that what makes a good scientist or artist, or a good intellectual in general, *by definition*, includes the characteristic of eschewing the temptation of seeking power for its own sake. Intelligent people who derive a satisfaction in doing things to help advance society, its knowledge, thought and practical state, do their best to avoid the everyday internecine power struggles of those who see power and influence as the means to social importance, and as an end in itself. Nevertheless, it is sufficient for this argument to recognise that the majority of ambitious people who seek to make a lasting mark on society belong to the second group. It is an unfortunate weakness, and maybe even a fundamental structural flaw, of humans and their societies that few rise to greatness, but many rise to power.

In the drive to achieve recognition and influence through power, individuals will use the mechanisms available to that end. It is not true that all people have an innate tendency to tyrannise. Some individuals will voluntarily relinquish the opportunity for power or influence over other human beings because they perceive it to be corrupted power. In this case, it is important to see that many people possess a sense of wider social responsibility alongside their drive for personal attainment, and this call to a wider responsibility provides an internal check on their behaviour. But the salient point is that collectively, because all human beings have different moral boundaries, there will always be somebody who is likely to exploit any opportunity to gain power and influence. It is the standard deviation in individual human moral proclivities that gives society, as a whole, a tendency to fill opportunities for tyranny, and therefore a tendency to tyrannise.

It follows from this principle that tyranny is likely expand to fit the moral, legal and practical opportunities that are offered to it. Unless active attempts are made to counter it by legislation and education, then the removal of the buffers to tyranny by the extremities of the extra-terrestrial environment, and the culture the removal of these buffers creates, will provide greater opportunities for despotism, and those opportunities will be seized. Implementing social checks against this tendency without undermining liberty itself will be one of the most profound challenges faced by extra-terrestrial societies from the moment of their creation.

14. The Liberation of Creativity

It is well known that a repressive environment may nurture some of the greatest intellectual achievements. The Stoic retreat impels a greater proportion of humanity to contemplation and, in instances where people are already inclined to this, to a more intense form of it. This is not an apology for tyranny, but if a repressive social order is imposed upon people by the harshness of the extra-terrestrial environment, may it not be the case that, in some of these environments, the political and social conditions that arise from the physical environment will nurture intellectual creativity?

Although it is likely that confronting the extremities of extra-terrestrial environments will eventually become the norm, and that technology to alleviate some of the threats will become so reliable until it is barely on the minds of people, there can be no technological fix for the raw reality of the instantaneous lethality of the environment. From this fact, there may emerge societies in which the repression and restriction imposed by the natural environment have been usurped and overcome by extraordinary intellectual originality and creativity in areas of human life that do not require such open and free physical movement: philosophy, art, science and ethics [47].

The predictability of extra-terrestrial environments will further drive people to seek new intellectual outlets and novelty.

Although most extra-terrestrial environments are extreme, many, particularly those of interest to people (the Moon, Mars, asteroids), are highly predictable on account of the absence of complex hydrological cycles and atmospheric circulation patterns. In the absence of a dust storm, for instance, the temperature experienced in any given place on the surface of Mars at any given time of the Martian year is tediously invariant from year to year. Danger only becomes a constant sink for intellectual energy when it is consistently unpredictable. Once an established population has successfully adapted to the physical environmental conditions, apart from the rare emergencies, they will be confronted by extraordinary banality in the day-to-day experiences of weather, climate and general environmental changes. An escape from this condition will be achieved through an intellectual efflorescence.

Each different extra-terrestrial environment will encourage intellectual thought that is pinned to both the physical nature of the environment and the historical context. For example, although the seeds of an environmental awareness are even to be found in ancient Greek philosophical thinking [48], there is nothing that approaches a true and fully formed environmental ethic. The explanation for this is that there was then no conception of the globally interconnected nature of the biosphere, and therefore no sense of any environmental crisis, or even of any potential for one. Only in the 20th Century does this insight emerge, and then there is an explosion of writings in environmental ethics [49]. Similarly, the concerns of Hume, Locke, Hobbes, Kant and other philosophers of the Enlightenment were about the nature of societies, and the extent of state power. These thinkers and their contemporaries occupied themselves at length with the role of religion, albeit cautiously. By the time J. S. Mill was examining society, the foundations of religion were much more acceptable topics for open discussion. Although each age has its philosophers, the breadth of their discussions is determined by the social topics of the day, their historical roots and the environment in which they thought. In extra-terrestrial environments, we can expect that thoughts will similarly be motivated by the specific issues that dominate at any given time, but we can also observe some general points on creativity that will probably apply.

The nurturing of inchoate, but highly novel, ideas in philosophy and science is facilitated by a lack of precedents [50]. Necessarily, all good ideas derive some of their inspiration from previous ideas, but an environment in which there is a saturation of ideas covering many different fields may stifle radical creativity. There is evidence for room for new radical ideas in many branches of philosophy and science on the Earth, but it must certainly be the case that when many of the traditional canons of thought we are familiar with on Earth have become distant to an extra-terrestrial population, that population will be intellectually freed by a recrudescence of many of the central ideas of society. They will read Kant and Schiller on electronic media of some kind, but these thoughts will seem somehow irrelevant to them, as neither of these philosophers or any of those who came immediately before or after them were interested in extra-terrestrial society (although scientists such as Huygens before, and Herschel afterwards, did certainly speculate on the presence of extra-terrestrial life). There will be a driving deeper urge to establish a new branch of philosophical thinking that is specific to them and their extra-terrestrial societies, thinking that they can claim as their own, and that more readily reflects their particular social predicament. Their thoughts will potentially be rich since they will have not only the distance of history between them and their forebears, but they will inhabit quite literally a different world. The opportunity this offers amounts to a type of 'intellectual release' from past traditions, an extra-terrestrial Renaissance if you will. Without the boundaries of terrestrial society to frown on innovative thinking as *outré*, or to make these intellectual excursions hidebound to the existing norms and customs of terrestrial society built upon 50,000 years of human history, a philosophical and scientific open plain [51] will open up for them to wander across unhindered, vivified by new experiments in thought and social organisation.

Perhaps then, and despite the picture I have painted, extra-terrestrial environments will in fact nurture a remarkable and intellectually productive period in human history, and will bring forward individuals with views and visions of society that are not only different from those that we know on the Earth, but also offer new and penetrating insights into the human condition, and may reinvigorate the intellectual quality of civilization in its widest sense.

Notes and references

1. Plato's *The Republic* is the earliest exposition of a 'perfect state'. The rigidity of its proposed social order (a strictly engineered caste system) and its conception of an exquisitely planned society has made it one of the most lauded attempts at devising a rational social politics, but equally it can be seen to be a picture of a despotic state in which freedom is all but extinguished, a point eloquently explored by Karl Popper (K. Popper, *"The Open Society and Its Enemies"*, Routledge and Kegan, 1945).
2. I use the word 'liberty' here to mean the desire for a minimum extra-terrestrial state and minimum intervention in social and economic matters and the private lives of people; in some sense it is consistent with a libertarian point of view (e.g. 'libertarianism stresses the ideal of civil society – meaning a society that eschews dealing with one another by subjugation, oppression, conquest or similar coercive means', T.R. Machan, in C. Duncan, T.R. Macahan, *"Libertarianism: For and Against"*, Rowman and Littlefield, p. 5, 2005). However, although I do not address this specific issue in this essay, I do not suggest a strong libertarian (i.e., anti-state) position, such as autarchism. There are two reasons: 1) some type of centralized structure is necessary, and desirable, to maintain safety in extra-terrestrial environments, despite the way in which this might be abused, which is the subject of this essay (although this state could still be a minimal state), 2) paradoxically, in extreme environments, the state will be the most effective way

to secure the liberation of the individual from the huge efforts required in life support to maintain human existence. This paradox applies to most people on the present-day Earth because many time-consuming activities, such as waste removal, are dealt with by national or local governments, freeing individuals from these tasks. But the strength of this paradox will be much greater in extra-terrestrial environments. On the Earth, it probably first emerged in early human societies when, for example, agrarian communities, in pursuing the central organization of food productivity, freed individuals from the need to hunt and gather their own food.

3. Of course, the attempt to design tyrannical tendencies out of a society is, in itself, a form of social engineering and could be viewed as a form of tyranny; but any society, particularly a democratic society, that attempts to protect freedom is in the process engaging in social engineering, which paradoxically represents the antithesis of the pursuit of freedom.

4. A further ambiguity in the phrase 'free society' lies in the observation that there is not necessarily an equivalence between the freedom of a whole society (a 'free society') and its constituent members (P. Van Parijs, *"Real Freedom For All: What (if anything) Can Justify Capitalism?"*, Clarendon Press, 1997). As Van Parijs points out, a society might have the freedom to defend itself against a foreign power, but if this involves conscription, its individual members may not be as free as a society that does not resort to conscription. In the case of extra-terrestrial societies, this ambiguity will be irrelevant until a time when separate settlements engage each other to create inter-society conflicts and agreements, such that the freedom of a society as a single entity becomes a meaningful concept (although in a looser way one could talk about the total general level of freedom exhibited in an isolated society compared to others, say, for example, on the Earth).

5. The paradox that one person's freedom will sometimes infringe another's is perhaps still best captured by historian Richard H. Tawney's overused quote: 'But freedom for the pike is death for the minnows', R.H. Tawney, *"Equality"*, Allen and Unwin, p.208, 1938, or the equally worn, and somewhat more crass maxim, 'Your freedom of action ends where my nose begins'. From a more philosophical standpoint, the 'paradox of freedom' also applies to the more abstract idea that we cannot escape being free (at least within our own minds). The point, explicated by Jean-Paul Sartre, is that we are 'condemned to be free'.

6. Many philosophers' work has been concerned with the conundrum that if the world is objective and ultimately predictable through scientific investigation (including our understanding of the functioning of the human brain), how can any human being be described as having 'free will' in a real sense? This question need not concern us here, since we can still proceed from the assumption that people have choices and coercion limits those choices regardless of whether the mental basis behind taking a choice is argued to derive from 'free will' or biochemical determinism.
7. See W.G. Sebald and A. Bell, *"On the Natural History of Destruction"*, Penguin Books, 2004.
8. G.W.F. Hegel, *"The Phenomenology of Mind"*, translated by J.B. Baillie, Harper & Row, 1967 (first published in 1910).
9. It is an obvious point, but worth underscoring, that humans are the result of over 3.5 billion years of terrestrial evolution and are therefore exquisitely evolved for living on the Earth. Any extra-terrestrial environment will, by default, be 'extreme'. The only exception to this would be an extrasolar planet orbiting a distant star that possesses a 21% oxygen atmosphere and an environment much like the Earth, allowing for a seamless integration (although the ethics of such an integration would be questionable – e.g., C.S. Cockell, "The ethical relevance of Earth-like extrasolar planets", *Environmental Ethics*, 28, pp. 303-314, 2006). It follows that the relative extremes in comparison to the Earth discussed in this essay are an ineluctable part of the creation of an extra-terrestrial society, although the types and combinations of extremes pressed upon societies will vary from location to location.
10. The production of rocket propellant using Martian resources is described by R.L. Ash, W.L. Dowler, G. Varsi, "Feasibility of rocket propellant production on Mars", *Acta Astronautica*, 5, pp. 705-724, 1978. This paper established a concept of 'In-Situ Resource Utilization (ISRU)' for Mars. Other studies have examined a diversity of ways in which natural resources on Mars can be used to sustain a human presence, e.g., T.R. Meyer, C.P. McKay, "Martian in-situ resource use", *JBIS*, 42, pp. 147-160, 1989. ISRU is also a central theme in R. Zubrin, *"The Case for Mars"*, Pocket Books, 1998. The diversity of industrial processes that could be developed on Mars is huge and schemes for the production of plastics, glass and other materials have been proposed in various journals and books. All of them, however, require the application of coordinated technological approaches and methods that are difficult for mere individuals to master without collective industrial support.

11. The analogy between Antarctica and Mars is a well-worn intellectual path. There are clear similarities which include the monotonous landscape, isolation and the inherent physical dangers in the environment (e.g., A. Harrison, Y.A. Clearwater and C.P. McKay, *"From Antarctica to Outer Space"*, Springer, 1991). However, Antarctica is still benign in many of its characteristics compared to most locations in outer space, not least the breathable atmosphere. By contrast, it should also be pointed out that Antarctica can expose people and infrastructure to devastating storms and 'white outs'. On the Moon there is no significant atmosphere and on Mars the low atmospheric pressure limits the damage that high winds can cause. This may be one of the few climatic factors in which the Moon and Mars are an improvement on Antarctica.

12. C.S. Cockell, "Mars is an awful place to live", *Interdisciplinary Science Reviews*, 27, pp. 32-38, 2002. I used the polar demographics of the terrestrial High Arctic to predict a maximum population on Mars of about 3 million people, but suggest that this is likely to be over optimistic. The population density of the Antarctic is used to yield a maximum Martian population of half a million.

13. That climate would be likely to have an influence on the development of civilization is an old idea; French philosopher Montesquieu made the most deliberate point of recognizing that climate is one important factor that governs the development of civilization. In some sense, we can regard discussions on the influence of climate on society as the philosophical forebears of the recognition of the likely importance of extra-terrestrial environmental conditions on future societies in space. Turgot also understood that climate would have some influence, but his historical account of the development of human society recognized that probably climate would be subordinate to the overprint of the culture and historical context of each society.

14. Contrasting styles of leadership exhibited by Scott and Amundsen have been a topic of controversy and historical revisionism, made more obscure by the myths surrounding the exploits of both men. An account that takes a dim view of Captain Scott is R. Huntford, *"Scott and Amundsen"*, Putman, 1981 and can be contrasted with R. Fiennes, *"Scott"*, Hodder and Stoughton, 2003. Regardless of one's view, the technological approach and focus on innovations employed by Scott and the approach relying on old native knowledge applied by Amundsen remains one of the most conspicuous differences in their styles of exploration.

15. ' The problem of anomie may grow into serious proportions on a national scale only in a relatively free country and in the absence of one acute specific danger that unifies the whole population in defense', C. Bay, *"The Structure of Freedom"*, Stanford University Press, p. 119, 1958. If the extra-terrestrial environment is the cause of anomie, it is a relatively easy task to use extreme environmental conditions to unite populations, if necessary through coercion, to counter it.
16. 'Basically what shaped its [kibbutz] character was the necessity for adaptation to the unusual conditions obtaining in Palestine. Hence, the peculiar social structure was necessary to ensure survival', H.F. Infield, *"Cooperative Living in Palestine"*, Dryden, p. 25, 1944.
17. The collective work ethic of a small society harbours an ambiguity in whether such an environment liberates individuals or enslaves them. For example, Spiro states 'another principle underlying the culture [of a kibbutz] is that of individual liberty; indeed, the kibbutz prides itself on being the freest society in the world', and then without the slightest sense of a possible contradiction, says one page later, 'It means, first, that the interests of the individual must be subordinate to the interests of the group', M.E. Spiro, *"Kibbutz: Venture in Utopia"*, Harvard University Press, p. 28 and 29, 1981. It is possible to write these two statements without a sense of contradiction if one believes that the egalitarian distribution of the products of individual labour, and thus subordination of individual ambitions to group work, liberates the individual from the competitiveness of a less egalitarian society. But the veracity of such a (Marxist) position is itself a matter of worthwhile debate. As one kibbutznick is paraphrased more recently, 'The traditional kibbutz assumed that if the community was successful, the individual would be secure. Samar puts it the other way round: if the individual is cared for, the community will be successful', D. Gavron, *"The Kibbutz: Awakening from Utopia"*, Rowman and Littlefield, p. 263, 2000. The tension caused in small societies by the conflicting interests of the group and the individual is evidenced in this potentially confused and self-contradictory quote from yet another young kibbutznick, 'Real equality is'...'equality within diversity, an equal right to self-realization within the collective egalitarian framework of the kibbutz', p. 186.
18. A complete withdrawal from society will not be possible. However, the desire to remove oneself partially from the social interdependence in these environments, together with the self-

control necessary to live a fulfilled life in these environments, is likely to nurture a proportion of individuals who exhibit a mixture of Stoic, Epicurean and Cynical characteristics (self-control, assertion of reason over emotion, virtue, self-sufficiency, etc.), at least seen from the classical characteristics of these philosophies.

19. In this essay I prefer to put despotism, dictatorship and totalitarianism into the same pot. A convincing argument can be made that there is in fact a distinction between dictatorship, which involves a well-planned hierarchy, and totalitarianism, which takes on the character of a large amorphous 'movement' supported by systematic terror (e.g. Nazi Germany and Stalinist Russia). In the sense that a despot can rule over a small group of people, but totalitarianism requires a large number of people (on the scale of nations) to generate the structures of mass control and terror, true totalitarianism is unlikely to emerge in an extra-terrestrial environment because the numbers of people will be, in most locations, too small. In most extra-terrestrial societies the worst form of organised control will be derivatives of dictatorship.

20. I. Berlin, "Two Concepts of Liberty", in I. Berlin, *"Four Essays on Liberty"*, Oxford University Press, pp. 118-172, 1988. Berlin provides a particularly clear discussion of the separation between negative liberty (freedom from interference) and positive liberty (freedom to do certain things), although the recognition of the separation between these two types of liberty precedes Berlin (e.g., C. Bay, *"The Structure of Freedom"*, p. 57 and M.J. Adler, *"The Idea of Freedom"*, Doubleday & Co., 1958). Insofar as the freedom to do a certain thing usually implies a lack of coercion against doing it (i.e., potential interference), then one can construct an argument that there is no difference between positive and negative versions of freedom, but in this essay I separate them according to the meaning of Berlin for convenience.

21. 'Real power begins where secrecy begins', H. Arendt, *"The Origins of Totalitarianism"*, Harcourt, p.403, 1985. Arendt later elaborates in detail why it is that in totalitarian states the police must necessarily become an instrument of control since through them the 'atomisation' of society into paranoid individuals protecting their own interests becomes possible – later it becomes inseparable from the very existence and continuity of the totalitarian state. The control of information is one part of the development of this culture of separating humans. It also leads to the inevitable, and bizarre, paradox that the development of mass collective will embodied by totalitarian societies is only made possible by first pushing people to extremes of individualism by separating them

from family and friends. From this point their isolation can be used to reunite them into a 'movement'. Seen from this point of view it is possible to see how even capitalist societies that encourage individualism and interpersonal competitiveness could fluidly make the transition to totalitarianism.
The abuse of knowledge in this way is not necessarily a clear transgression of trust: 'There is probably no social interaction that is going on that does not include elements of manipulation. Most communication, if not all, includes conscious or unconscious attempts at regulating the supply of information in the interest of encouraging or discouraging certain types of behaviour', C. Bay, *"The Structure of Freedom"*, p. 320.
The link between knowledge and power has perhaps been most comprehensively investigated by Foucault, e.g., M. Foucault, *"Power/Knowledge"*, Pantheon Books, 1980.

22. 'The acts of an individual may be hurtful to others, or wanting in due consideration of their welfare, without going the length of violating any of their constituted rights. The offender may then be justly punished by opinion, though not by law', J.S. Mill, *"On Liberty"*, Oxford University Press, p. 83, 1988. This notion is dangerous because it is open to the obvious abuse, adumbrated in this essay, of individuals being singled out for behaviours or critique that is perfectly legal, but happens not to be convenient for authorities in power. Encouraging collective disapprobation by society is the most insidious instrument for subverting independence of mind and strengthening conformity.
There is good evidence that in small isolated populations public disgrace is a powerful, and regarded as a highly effective, tool of group coercion. Concerning the Inuit: 'Means of enforcing peace and harmony within communities included shaming, shunning, banishment, abandonment leading to death, and execution. Public ridicule and ostracism were the most frequently used methods of social control, and generally had the desired effect of keeping people cooperative', quoted in R. Fossett, *"In Order to Live Untroubled: Inuit of the Central Arctic, 1550-1940"*, University of Manitoba Press, p. 207, 2001. And in the Israeli kibbutzim: 'Should a chaver [Hebrew for friend or companion – used to refer to other members of the kibbutz] violate the group norms, it would not only be known in a short time, but he would be openly criticized for his behaviour. Such group censure, informal though it is, is highly effective' (p.99)….. 'The first procedure is to bring the person's dereliction to the official attention of the kibbutz at a town meeting. This is a powerful sanction, and the very threat

to use it is usually efficacious in this community where people are so sensitive to public opinion', M.E. Spiro, *"Kibbutz: Venture in Utopia"*, p. 100.

23. The attraction of actively constructing social orders that are designed to increase individual liberty is the product of the 18th Century French revolutionary vision of liberalism. However, from Plato onwards, the romanticism of 'designed' free societies has a long history. The end point of such designer societies is likely to be anything but an enhancement of individual liberty (e.g., I. Berlin, *"Freedom and its Betrayal: Six Enemies of Human Liberty"*, Princeton University Press, 2002.)

24. 'Necessity is the plea for every infringement of human freedom. It is the argument of tyrants; it is the creed of slaves', William Pitt the Younger, *House of Commons, London*, November 18, 1783; and 'If either a public officer or anyone else saw a person attempting to cross a bridge which had been ascertained to be unsafe....they might seize him and turn him back, without any real infringement of his liberty.... Nevertheless, when there is not a certainty... no-one but the person himself can judge of the sufficiency of the motive', J.S. Mill, *"On Liberty"*, Oxford University Press, p. 107, 1998. This view is defensible, but the unsafe bridge can equally become a convenient excuse for preventing people from crossing any bridges or building bridges to protect their own safety and the safety of others, and it is this motive that most profoundly threatens liberty in extra-terrestrial environments.

25. For example, 'That principle is, that the sole end for which mankind are warranted, individually or collectively, in interfering with the liberty of action of any of their number, is self-protection', J.S. Mill, *"On Liberty"*, Oxford University Press, p. 14, 1998. Many later writers do not disagree, but attempt to define the extent to which self-defence is justified, for example 'defensive violence, therefore, must be confined to resisting invasive acts against person or property', M.N. Rothbard, *"The Ethics of Liberty"*, New York University Press, p. 77, 2002.

26. Hugo Grotius used his concept of 'natural laws' to advance propositions for a just war, in which basic rules applicable to all warring parties would apply, independent of local customs and traditions, and established by 'natural' laws, H. Grotius, *"On the Laws of War and Peace"*, Kessinger Publishing, 2004 (first published in 1625).

27. T. Hobbes, *"Leviathan"*, Oxford University Press, 1998 (first published in 1651).

28. 'Nothing proved easier to destroy than the privacy and private morality of people who thought of nothing but safeguarding their private lives', H. Arendt in *"The Origins of Totalitarianism"*, p. 338 describing the influence of Himmler and his activities on the development of Nazi society. The recognition that humans can subordinate their concerns about a political system bent on genocide to their own personal interests was one of the most alarming and extraordinary facts about human nature that totalitarian societies brought sharply to light in the twentieth century (although it was certainly not the first time this possibility was realised practically).
29. Hobbes had a distinctly negative view of humanity as engaged in a perpetual war against itself, in contradistinction to either Rousseau, whose original state of nature was envisaged as humans isolated from one another and rarely engaging in social interaction until social bonds (for the example, the birth of agriculture) made a social contract necessary for expediency, or Locke, who had a more positive view of human beneficence and co-operation in this primitive stage. The lethal threat of the extra-terrestrial environment may pitch people into a Hobbesian state of paranoia about others' intentions. However, it might also encourage elements of a social contract that express benevolence and selflessness towards other humans as a necessary means both to ensure survival and live peacefully.

The 17th and 18th Century fascination with defining the original 'state of nature' was almost certainly brought about by a lack of understanding of human evolution and its links to the evolution of other primates. It was only natural, in the absence of this information, for curious intellectuals to try to define what the 'original' human society must have looked like and how its members might have behaved. In reality there was unlikely to have been an original state of nature, but instead a seamless gradation of behaviour from apes to early hominids. From this perspective Hobbes's war of all against all, and Locke's essentially benevolent people can be seen as two end member types of behaviour which humans express in varying degrees at any given time. Probably the least realistic view of early societies was Rousseau's (J-J. Rousseau, *"The Social Contract"*, Penguin, 1976 (first published in 1762)) because humans are not fundamentally solitary creatures, although incorporating solitary behaviour and our 'private lives' within the 'social contract' of collective social organisation, remains, as he recognised, one of the major challenges to the assembly of peaceful human societies.

30. Numerous small isolated societies might yield insights into the collective patterns of social organization that could emerge in extra-terrestrial societies. Some isolated groups, such as those on drilling rigs or in nuclear submarines, are not long-term societies and so offer few lessons in the development of long-term political and economic systems. The Israeli kibbutz is one such permanent society (H. Barkai, "The Kibbutz: An Experiment in Micro-Socialism", in I. Howe and C. Gershman, *"Israel, the Arabs, and the Middle East"*, Bantam Books, pp. 69-99, 1972) which offers an insight into the evolution of a socialist egalitarian structure. Many kibbutzim have, since the middle of the 1990s, transformed their economic behaviour towards capitalism, suggesting that no system is 'natural' and that, at least on the Earth, both socialist and capitalist economic models can be adopted by the same isolated groups of people. These transitions show that it is difficult to predict meaningfully the political and economic trajectory of an isolated extra-terrestrial group, and that a single extra-terrestrial environment could in fact host diverse models of development. The deep social bonds found in the Inuit of the High Arctic are yet another example. The Inuit offer a particularly interesting model of isolated societies with a long history because they are exposed to extreme polar (in some respects Mars-like) conditions. All of these models, however, suffer from two weaknesses: 1) none of these societies is exposed to the collected extremities found in extra-terrestrial environments, in which the peculiar political and economic style of human organization ultimately can only be determined by actually living in these environments, 2) the transition to modern technical societies in these groups has occurred through the process of external influence, in the case of the kibbutzim, primarily from the Israeli state and society. Extra-terrestrial societies will start with a high level of technical competence, and although they will initially receive influence from the Earth, their extreme isolation means that their development, internally and in relation to external influence, will not exactly recapitulate isolated groups on the Earth.
31. The development of extra-terrestrial society may in some ways recapitulate the history of human civilization on the Earth, from small early tribal networks to nation-states, as the scale and complexity of extra-terrestrial infrastructure increases. Just two of the many brilliant expositions of the nature of liberty and its correlation with the state of human civilization in different stages of development can be found in R. Muir, *"Civilization and Liberty"*, Jonathan Cape, 1940 and the earlier and more remarkable,

given the circumstances under which it was written, Cordorcet, *"Sketch for a Historical Picture of the Progress of the Human Mind"*, Weidenfeld & Nicolson, 1955 (first published 1795). There are two obvious distinctions to outer space, however: 1) extra-terrestrial societies may be comparatively a blank social canvas since detached from the Earth they may have more freedom to attempt new experiments in social organization, 2) they will have access to the accumulated technological prowess of human society on Earth and will begin these 'early' stages of civilization with great scientific and technical insight.

32. My analysis here comes closest to the points on oligarchy made by Michels (R. Michels, *"Political Parties: A Sociological Study of the Oligarchical Tendencies of Modern Democracy"*, Free Press, New York, 1962). Michels' analysis investigates in depth the effects of oligarchy in a range of organisations. Here I am principally concerned with how these tendencies would modify organisations in the space environment.

33. 'Is one therefore to conclude that economic centralization aiming at social security must sweep away spiritual freedom?', J.L. Talmon, *"The Origins of Totalitarian Democracy"*, Frederick A. Praeger, p. 255, 1961. The relationship between economic and political freedom is ambiguous. Friedman states 'economic freedom is also an indispensable means toward the achievement of political freedom', M. Friedman, *"Capitalism and Freedom"*, University of Chicago Press, p. 8, 2002 (first published in 1962), and in the updated Preface in the 2002 edition, states that 'political freedom, although desirable, is not a necessary condition of economic freedom' (ix). Hayek has little doubt that a centrally planned economy, simply from the necessity for central organization, will erode democracy, F.A. Hayek, "Economic Control and Totalitarianism". in *"The Road to Serfdom"*, Routledge, pp. 91-104, 2007 (first published in 1944). It is not clear how political and economic freedoms can or will be linked in extra-terrestrial societies. Indeed, if the enclosed spaces of extra-terrestrial environments are places where corporations find fertile ground for cartels and wealth accumulation at the expense of an enslaved population, then the centralized control of the distribution of wealth may even be desirable in some situations.

34. The character of capitalism itself may be contingent on historical circumstances (M. Weber, *"The Protestant Ethic and the Spirit of Capitalism"*, Routledge, 2007 (first published in 1904)). It is likely that given the need to survive extreme environmental conditions, the puritan sense of 'duty' would be manifested strongly in extra-

terrestrial capitalism, but that the undercurrent of collective social responsibility caused by these environments might dampen the drive for individualistic acquisition of wealth for its own sake, as exemplified by small communities on the Earth. Any attempt at extra-terrestrial capitalism could easily end in a type of 'subdued' capitalism underpinned by a strong collective work ethic.

35. 'Everywhere do I perceive a certain conspiracy of rich men seeking their own advantage under the name and pretext of the commonwealth', the words of Sir Thomas More, quoted by J.A.Hobson, *"Imperialism: A Study"*, Cosmo Classics, 2005, p. 46 (first published in 1902).

36. It is disingenuous to attack hard working civil servants who work for the benefit of their country and its population, but it remains an irrefragable fact that this branch of society has the most administratively useful potential to an emerging tyranny.

37. I. Carter, *"A Measure of Freedom"*, Oxford University Press, p. 29, 1999.

38. Mars experiences both volcanism and earthquakes. However, neither occurs, at least in the present-day, with the devastating consequences seen on the Earth. It is a strange observation that freedom is connected to plate tectonics, but in certain realms of life it is. Mars is comparatively free of the physical dangers associated with moving plates compared to the Earth, and consequently the specific systems of administration, oversight, warning, and evacuation associated with earthquake and volcano-prone regions of the Earth will not be required. However, this release from the politics resulting from one form of extremity will be overshadowed by the systems of administration associated with the lethal atmosphere.

39. 'The world operates a certain way, according to causal laws, and the constraints imposed by nature are the foundation for human choice, not a barrier to it', D. Kelley, *"A Life of One's Own: Individual Rights and the Welfare State"*, Cato Institute, Washington, p.69, 1998; Philippe Van Parijs incorporates the recognition that natural barriers to freedom cannot truly be recognized as restrictions to liberty in his concept of 'real freedom', P. Van Parijs, *"Real Freedom For All"*. Real freedom requires that the individual actually has the capacity or resources to carry out their will. From this perspective extreme environments, on Earth or in space, although they restrict many freedoms such as movement, cannot be considered as real restrictions to freedom because no political or personal will can overcome them – they are simply realities of the environmental conditions in which one lives. Nevertheless, these physical

conditions can still cause restrictions in political and social freedoms, either by directly restricting people and their ability to move, for instance, or encouraging restrictive social policies and practices. The link between the physical environment and restrictive social and personal freedoms they create may become so intimate that in outer space Parijs's distinction becomes moot – individuals will consider the outside physical environment, no matter how unchangeable, to be a contributory architect of tyranny.

This distinction does not mean, of course, that we cannot expand our opportunities for social and political freedom by overcoming natural or physical restrictions – 'We may give the name 'physical freedom' to the mastery over non-human obstacles to the realization of our desires', B. Russell, "Freedom and Government", in R.N. Anshen, *"Freedom – Its Meaning"*, George Allen and Unwin, p. 231, 1942.

We can even use physical phenomena that usually threaten liberty to expand it, for example Malinowski's perceptive comment, 'The contribution of fire to freedom consists in that it extends the range of human action', B. Malinowski, *"Freedom and Civilization"*, Indiana University Press, p. 113, 1964 (first published in 1944). But the inability to overcome physical barriers does not constitute a *denial* of liberty.

40. Weapons of mass destruction are prohibited in outer space by United Nations agreement, but as we witness the difficulties of controlling the proliferation of these weapons on Earth, how are we to police their spread in outer space? The number of asteroids in our Solar System greater than 1 km in diameter on which a chemical or biological weapons factory could be hidden, or a nuclear weapon stored, is estimated to be much greater than 1.1 million. Ultimately colonies in outer space, particularly those on planetary surfaces, isolated and at the mercy of malevolent individuals and groups hidden in the vast tracts of interplanetary and interstellar space, may be more prone to devastation at the hands of other human societies than we have witnessed on the Earth. The threat of annihilation in the infinite un-policeable expanses of space may itself yield a quite unique culture of suspicion and paranoia which would fuel a culture of control.

41. I. Brant, *"The Bill of Rights: Its Origin and Meaning"*, Bobbs-Merrill, 1965.

42. Friedman states, on parental responsibility and children: 'But reality has its own discipline. The alternative to parental authority is and should be freedom…Experiencing the real world directly

– learning to survive in it – is not as pleasant a way of growing up as being taught about it by one's parents. But if the parents are unwilling or unable to do the job, it may be the best substitute available', D. Friedman, *"The Machinery of Freedom"*, Open Court, p. 94, 1995. On a planet where the behaviour of miscreants, and even the physical damage they may choose to cause, can be absorbed into society, this solution is attractive. But it is a luxury that a small society surrounded by an instantaneously lethal atmosphere can ill afford to have when errant conduct is more likely to be fatal.

43. The extremity of extra-terrestrial environments and the difficulty of moving around in them coupled with the lack of the diversity of goods compared with the Earth (at least in the early stages of extra-terrestrial settlements) will nurture a form of asceticism in social and individual conduct. Many individuals may, if they are allowed to, combat this social environment by deliberate attempts to expand their social freedom and their acquisition of goods, but these environments will provide a stage for the flourishing of ascetic religious beliefs as a means to come to terms with the environment – almost a form of religious resignation.

44. J. Locke, *"Two Treatises of Government"*, Everyman Library, 1993 (first published in 1689). Locke's view that land can become private property by working on it was suggested as an approach to encourage people to settle otherwise barren and unattractive extra-terrestrial environments, since the possibility of converting land into one's own private property is one of the few tangible incentives that can drive people to settle desolate environments (C.S.Cockell, "A Simple Land Use Policy for Mars", in J.D.A.Clarke, *"Mars Analog Research"*, American Astronautical Association, 111, pp. 301-311, 2006). Bertrand Russell correctly pointed out that Locke's ideas have lost meaning because corporations and individuals can claim ownership of land without actually working on it themselves (B. Russell, *"History of Western Philosophy"*, Touchstone, p. 636, 1967). They can even sell it on without using it for anything productive at all. However, we might wonder whether, as a general policy, Locke's ideas might have some value in outer space. If individuals or corporations were required to work on land in a very real physical sense before being able to acquire it as property, it would prevent a solar system-wide armchair 'land grab', and force organizations to expend the effort to develop human settlement in space prior to their ability to claim land. In addition, the principle that once people have worked on the land, then they have the right to claim it, would provide a strong incentive to explore and settle space.

45. 'Planetary protection' embodies the concerns about the contamination of other planets either hosting an indigenous biota or capable of hosting one ('forward' contamination) and the concern about the infection of the Earth's biota with a pathogenic or potentially environmentally pervasive extra-terrestrial organism ('back' contamination). It is not international law at the time of writing, but the degree of allowable contamination on spacecraft destined for planetary environments that might have a chance to harbour life is defined in policies established by international agreement; see, for example, A. Debus and M. Viso, "Planetary protection main requirements for planetary environment preservation and life detection experimentation", *Acta Astronautica*, 51, pp.1-9, 2002; J.D. Rummel, P.D. Stabekis, D.L. De Vincenzi and J.B. Barengoltz, "Cospar's planetary protection policy: a consolidated draft", *Adv. Space Res.*, 30, pp. 1567-1571, 2002; C.P.McKay and W.L. Davis, "Planetary protection issues in advance of human exploration of Mars", *Adv. Space Res.*, 9, pp. 197-202, 1989. Also see, 'Treaty on principles governing the activities of States in the exploration and use of outer space, including the moon and other celestial bodies', UN Resolution 2222 (XXI) 1967, a formal international agreement governing conduct in space.
46. Perhaps the most notable early assertion that social interactions are motivated by human vices (particularly self-interest and greed) was developed by Bernard Mandeville in his satire *"The Fable of the Bees"*, Liberty Fund, 1988 (first published in 1705), although Mandeville went further and proposed that many of these vices lead to benefits to the general public (hence the subtitle of his book *'Private Vices, Publick Benefits'*). I am not suggesting here, however, that the drive for power is necessarily motivated entirely by self-interest or greed, but my thesis here is that it primarily stems from a personal desire to achieve something significant – itself motivated by a fear of mortality and insignificance.
47. 'Humiliating to human pride as it may be, we must recognize that the advance and even the preservation of civilization are dependent upon a maximum of opportunity for accidents to happen', F.A. Hayek, *"The Constitution of Liberty"*, Routledge, p. 27, 2006.
48. A developed discussion on the relationship of ancient Greeks to their environment and how that relationship was fashioned by the major philosophical schools is found in L. Westra, T.M. Robinson, *"The Greeks and the Environment"*, Rowman and Littlefield, 1997. Aside from the fact that there was no globally-encompassing environmental 'crisis' for ancient Greeks to recognise,

fundamental philosophical tenets unpinned a very different view of Nature, not least the ancient Greek view that things were unchangeable and fixed, challenged only by Heraclitus's assertion that everything is in flux. In a world where the environment's destiny is either fixed or controlled by god(s), there is little reason for humans to care, or even need to care, about their insignificant impact. In that sense early Greek and later European societies, even beyond the Enlightenment, despite their imperial ambitions, were too humble and modest in their assessment of the power of human industrialization to recognize its ultimate possible impact on the environment.

49. The journal, *Environmental Ethics*, has become one of the primary means of scholarly discussion on the ethics of our treatment of the environment, although the number of books on environmental ethics is also very large. In the US, forester Aldo Leopold is generally recognized to be one of the founders of this awakening (A. Leopold, *"A Sand County Almanac"*, Oxford University Press, 1949), although it is equally likely that it was as much caused by a simple pragmatic recognition of the environmental damage being done by the sheer scale of human industrial activity manifested as insecticide poisoning, forest clearing, etc.

50. 'It is wherever man reaches beyond his present self, where the new emerges and assessment lies in the future, that liberty ultimately shows its value', F.A. Hayek, *"The Constitution of Liberty"*, p. 340.

51. The 'open plain' of intellectual adventurism in extra-terrestrial environments will necessarily still be bounded by what is practically possible in outer space. A pertinent example is Nozick's minimal or 'night-watchman' state (R. Nozick, *"Anarchy, State, and Utopia"*, Blackwell, 2003). The minimum apparatus required to construct a state will be defined by the minimum set of systems required to ensure individual safety and the implementation of commonly agreed laws. However, if the errant behaviour of one individual has the potential to threaten many more people than on the Earth, then the minimum extra-terrestrial state would necessarily be larger because the state apparatus necessary to minimise the chances of criminality must be more extensive to achieve a correspondingly equal chance of effective enforcement. Thus, although extra-terrestrial environments will yield new opportunities for thoughts on social organization, they will at the same time be ring-fenced by the realities of the environment, and some conceptions of the state may be more restrictive than on the Earth.

Liberty and the State

I felt dissatisfied with the previous essay because I had discussed liberty, but made few suggestions on how institutional arrangements might be engineered to maximise its chances of preservation and success in space. This essay was an attempt to take the ideas I elaborated in the previous essay and suggest what the implications might be for the institutions and practices of the extra-terrestrial state. It was not my intention to go into great detail about how the state might be constructed and ordered, but instead to consider the general and broad implications of the pursuit of freedom in space to the type of extra-terrestrial state that might be optimal.

Liberty and the Limits to the Extra-Terrestrial State

The physical conditions that inhere in extra-terrestrial environments have a tendency to drive society toward collectivist mechanisms of political and economic order to successfully cope with, and prevent possible disaster caused by, the lethal external conditions. Liberty will therefore be eroded by deliberate human action, through extra-terrestrial authorities, and through a natural restriction in concepts of liberty that will attend the development and behaviour of people in confined environments. The encouragement of extra-terrestrial governance that nurtures liberty in outer space will require the formation of many institutions that encourage competition and reduce political and economic monopolies – with the legal system to sustain them. This problem is most clearly manifest in oxygen production. These considerations allow the purpose and limits of the extra-terrestrial state and precursor forms of governance to be circumscribed. Far from being a purely speculative enquiry, this discussion allows requirements in physical architecture and social organisation to be identified that can be considered from the earliest stages of space exploration and settlement.

1. Introduction

In which institutions in outer space should people put their trust for governance? What types of institutions will be the purveyors of tyranny and which institutions will provide the greatest bulwark behind which liberty can successfully flourish? It is presumptuous to try to design a state. When attempts are made to do this, the result is usually a utopian vision that is intellectually intriguing, but of limited practical utility [1]. Therefore, there is no value in seeking to provide an all-encompassing design of an extra-terrestrial state, laid out in the way that an architect might draw the plans for an entirely new city [2]. Nevertheless, it is possible to say something about the generalities of institutions, and sometimes specific institutions, that should be created to maximise freedom within the likely general conditions for liberty in outer space.

An essential pre-condition for suggesting how a state and its precursor forms of governance should be ordered is to understand first the way in which liberty might be expressed in extra-terrestrial environments. The major threats to liberty come from deliberate and direct human action – the tendency of individuals to tyrannise others when they are offered the opportunities to do so, which is magnified in outer space; and the natural tendency of the environment to create the conditions for tyrannical institutions and people [3]. Both of these opportunities spring from the lethal conditions to be found in any environment in outer space and the effect these conditions have on political and economic institutions and their constituent members.

Outer space is a concatenation of extremes that are rarely found anywhere, in the same combination, on the Earth. On all planetary bodies known to date there is a lack of liquid water that can be easily acquired by individuals. This is not unique to extra-terrestrial environments. Bedouin who subsist in terrestrial deserts make do without large quantities of water, but in space this condition is without reprieve and there are no oases that exist naturally, free of technological origin. There is also the problem of growing food. It is difficult to grow food in extra-terrestrial environments, but many regions of the Earth are not natural oases for vegetation either – food cannot be easily grown free of greenhouses and technology in the High Arctic or in Antarctica [4]. In extra-terrestrial environments, there is no possibility of growing any food in the natural environment without technological support.

To these two extremes is added a third [5]: the planetary-wide lethality of atmospheres or the lack of them. The atmospheric conditions on other planets, or the lack of atmospheres, are the primary cause of the previous two extremes discussed. Surface conditions on other planetary bodies or the absence of gas in outer space are inimical to biological material, and therefore to crops and animals associated with a human presence. Atmospheric conditions on all planetary bodies within our Solar System are not conducive to the stability of liquid water, although water may be found under thick ice crusts – such is the case in the Jovian moon, Europa.

The fatal conditions on other planetary bodies and in outer space, coupled with the difficulties that arise in water and food availability, are the conduit through which tyranny will develop. People can subsist without water and food for many days, but of course they cannot survive without breathing for more than a few minutes. In environments where breathing requires the support of a complex industrial process, then even the slightest control over any part of this infrastructure offers an unimaginable lever of control. This is a mechanism by which liberty can be eroded deliberately by people and the institutions they create.

In order for tyrannical influences to emerge it is not necessary that people deliberately go about using an extreme environment as an excuse for control. The environment will, almost imperceptibly, exert a despotic influence over people's minds, which results from the powerful incentive for collective action in the interests of safety. The degree of interdependence necessary to make an extra-terrestrial society run efficiently will stifle individualism and reduce the number of individuals willing to challenge their compatriots and the authorities that govern them or determine the laws and regulations under which they live.

Liberty-seeking institutions of the extra-terrestrial state can only be formulated or consistently defended when the specific threats to them have been identified. The two challenges that have been identified: deliberate action and unseen acquiescence toward tyranny–and their specific character in extra-terrestrial environments–provide the basis from which to understand the types of institutional structures that could be created to halt or minimise them. But before enquiring as to what these institutional structures might be, there is a fundamental question that must be briefly addressed: is there any justification for an extra-terrestrial state [6]?

2. The Justification for the Extra-Terrestrial State

From what historical precedent or foundations of political philosophy can the extra-terrestrial state and its precursor organisations claim their legitimacy? The canonical means to address this question on Earth is to consider the character of people in a 'state of nature', prior to any formal state structure, and then enquire as to how a state structure may be preferable, if indeed it is, to this state of nature [7]. The state of nature is probably too simplified a conception since a reasonably well-defined social hierarchy and order has probably been a defining characteristic of humans since their early evolution, as it is with other primates. Enlightenment philosophers conceived of this idea prior to knowledge of non-human primate behaviour and the evolutionary relationship between humans and other primates. It was therefore not surprising that they attempted to envisage some type of early human society without social order and that these visions of the state of nature were rather arbitrary and perhaps even unrealistic. Nevertheless, a state of nature, a type of anarchic primitive social order, is a useful hypothetical construct to try to understand what advantages an overarching organisational structure, or organised state, offers – in this specific case in the extra-terrestrial environment [8].

It is obvious that all extra-terrestrial states will be founded by states from the Earth and their respective space programs or corporations that emanate from those states. Even the most rapacious free enterprise extra-terrestrial society will be constructed by corporations whose legality has emerged from prior state structures, either on the Earth or in space. So unlike the vision of humans wandering barefoot and with the minimum means of subsistence, which has been a popular foundation of early philosophers' conceptions of the state of nature on the Earth, this type of early hominid condition will be unrealistic for early societies in space. Despite the fact that an extra-terrestrial state of nature in a practical sense could not even closely resemble the state of nature first posited by Enlightenment philosophers, and would in fact be very advanced, we could still ask the question: what would, hypothetically, the most primitive state of nature be like in an extra-terrestrial environment if these societies started from scratch without a state, similar to the hypothetical early states of nature that have been proposed for societies on the Earth?

The lethal environmental conditions in any extra-terrestrial environment make it impossible that individuals can exist on their own in the manner presumed by Rousseau, as solitary individuals. The most basic means of survival, for example a spacesuit, implies an industrial edifice that supports the individual. Spacesuits must have gas supplies, visors, layers of different materials to protect from the extreme environment, electronic systems and so on and so forth, all of which must be fabricated by an industry, probably many industries. The same is true of any other means of basic survival in other types of apparatus that protect the individual from the environment, for example a pressurised habitat. In addition, complex manufacturing processes are essential for other items required for life including the extraction of water from the natural environment and the growth of food. No single individual has the wherewithal to fabricate or collect all of these commodities on their own initiative. Assuming that the manufacturers involved in these industries are guided by some legal requirements that either pertain to the operation of the organisation or the individuals within it, then the presence of an industry that supplies the basic means of survival probably implies the presence of a state, however rudimentary and small it may be.

In contrast to Rousseau's solitary individuals we might imagine a state of nature with individuals interacting, but engaged in a war of all against all as envisaged by Hobbes. However, they will still compete with each other for basic survival needs in habitats or spacesuits. These conflicts will result in certain catastrophe in the confines of outer space. Even if we introduced the means of punishment to prevent these conflicts in a manner consistent with the state of nature elaborated by Locke, this does not change the requirement for forms of pressurised enclosure. No state of nature can avoid the requirement that there must be an industrial infrastructure to provide the whole array of subsistence items that humans would need to be in the state of nature in an extra-terrestrial environment in the first place. From this simple thought process emerges an elementary fact – the very act of being alive in outer space implies explicit consent to the state. This is not 'tacit consent' in the sense of Locke; to able to live in space is a quite overt acceptance of the infrastructure that allows one to be alive. Therefore there can be no state of nature, even a hypothetical state of nature, in outer space. To be alive is to accept the social contract and therefore, given the vast technical empire required to sustain human life and the legal systems under which it must be overseen, the state. Death is the only escape from the extra-terrestrial social contract. Seen from a more fundamental level, the industrial process required to gather oxygen creates an unassailable link between the act of breathing and the legitimacy of the extra-terrestrial state.

Unfortunately, no reprieve from this conclusion can be found by rejecting the social contract theory and instead pursuing a utilitarian justification for the state, which seeks to find support for the state in evidence that it increases the individual or aggregate happiness of the public. It would be a comic understatement to assert that to have a pressurised habitat or spacesuit makes people happier than if they lacked such equipment. If they did not have these basic requirements for survival they would very quickly die. If we assume that the industry to produce these commodities must be guided by some general laws developed by a state, then the utilitarian defence of the state rests on the foundation that it facilitates the production of commodities that not merely make people happier, but allow them to survive at all. In this context, the traditional arguments about whether this utilitarian advantage is accomplished by making individuals happy or the aggregate public become moot.

Although the utilitarian justification for the state, like the social contract theory, seems a watertight case, one might also note that the state derives a powerful justification in its capacity to reduce criminality. As the damage inflicted by gangs of anarchic criminals in enclosed pressurised spaces is potentially catastrophic, the state, as it does on the Earth, derives justification in its role as the arbiter and distributor of legally endorsed punishment which the public will demand to protect themselves.

The legitimacy of the state in extra-terrestrial societies originates in the requirement to establish the overall laws which will govern individuals and the corporations or state industries that will produce the means of survival, and the requirement to establish laws which will control criminality. The major dispute, therefore, does not concern the justification for the state, but how the state should be managed and what type of state it should be.

3. The Problem of Oxygen

It might seem illogical to begin with an examination of the major economical problem of freedom in outer space, since the reader might feel more inclined to suggest that political freedom is more important; but it seems that the supply of materials and the consumables upon which individual lives depend will create the foundation from which civil and political liberty will emerge. If resources are completely controlled by a central authority, then there can be little scope for political freedom in extra-terrestrial societies. This assumption may not be so apparent on the Earth, where the link between political and economic freedoms has been a subject of justifiable and intense debate. But, for reasons that will become apparent, it is the resolution of the economic problem in outer space which must be achieved first and which will be a central plank of the wider system of political freedom.

Each and every economic transaction lends itself to manipulation and abuse by the provider in the transaction, provided that those receiving the service or product are not aware that they could get a better deal or the product or service at a cheaper price. Where these transactions are localised to some specific economic activity or industry, they are primarily the concern of those taking part in the transaction. However, a serious social problem occurs where they begin to form a culture of a society and drive it towards a generally venal frame of mind; or where they begin to infringe on political and civil liberties where a product or service is vital to the health and welfare of people, who become subject to the economic ramifications of nefarious wheeler-dealings.

The extent to which this is a problem magnifies to immense proportions when it concerns a commodity on which the lives of people depend. The more vital a product is, the more dangerous monopolisation, price fixing and uncontrolled cartels become, but paradoxically, the more attractive the product becomes to those people inclined to this type of behaviour because of the enormous power vested in its essentialness to society. I cannot think of a single product on the Earth that is absolutely required, every second of the day, by all people, and that must be acquired by commercial transaction. If such a product existed it would be a product around which most economic activity revolved and which would become the intense focus of political and civil attention, agreements and disagreements.

For government officials this imaginary product would be a holy grail, for its control would not only secure power, but the longevity of the political system, since the population would be inclined to support a system that controlled something on which their lives depended. For private entrepreneurs, it would represent a hugely influential bargaining chip for commercial transactions. For private companies seeking to influence political decisions and economic policy in favour of their own interests, the control of even part of the market share of such a product would give them leverage over the unseen corners of political power, influencing their ability to sell even the most trivial products and manipulate policy to their own interests. For individuals seeking a place at the high table of power in society and a life of prestige, the control of this product or part of its manufacture would be a doorway to the highest echelons of political and social influence. So for both private individuals and the state the exertion of control over such a product must be irresistible.

For people on the Earth this vision is not a real one. There is no product of this type; even the most vital products we require, such as different types of food, can be acquired in many places from the natural environment, limiting the covetousness that is shown toward these resources. Unfortunately, such a product – oxygen – does exist on the space frontier.

The problem of oxygen inheres in the simple fact that it must be produced by a machine, which generates an entire industrial edifice through which the tentacles of tyranny can spread. The machine is likely to be built by somebody other than the person breathing the oxygen that emanates from it. If such a machine is centrally run and connected through pipes and tubes to living quarters, then an industry dedicated to maintaining the machine and its conduits of supply must be run by an organisation, whose internal workings may be obscure to the people breathing the oxygen it produces [9].

The arguments I elaborate here apply to the provision of water and food. They also apply to the systems that scrub carbon dioxide from the air. The removal of poisonous waste gases such as carbon dioxide is as vital as the provision of oxygen to human survival. However, I focus on oxygen. The requirement for this commodity on a second-to-second basis to sustain human life makes it a particularly lucid example of the potential mechanisms through which liberty can be threatened.

We can try to consider an optimised economic plan to minimise control. For example, imagine that such a situation was mitigated by each and every individual or family owning their own oxygen-producing machine. However, in this situation some organisation must supply the raw material for the machine to make oxygen. On the surface of Mars, carbon dioxide from the atmosphere can be cracked to produce oxygen, which could conceivably provide some measure of autonomy from a large organisation since the carbon dioxide could be drawn in and cracked by the machine itself. (Such an idealised situation would not exist on any other planetary body though. On the Moon, oxygen must be extracted as a raw material from rocks or possibly even polar crater ices). Given a perfect arrangement of individually owned machines using freely available atmosphere as a raw material, the machine still needs maintenance using tools likely to be produced by somebody other than the individuals breathing the oxygen. Organisations must exist to provide the spare parts and servicing. In the idealised extra-terrestrial society where all efforts have been made to achieve decentralisation of oxygen production, nothing can change the basic fact that technology of some kind is required to transform an atmospheric constituent into a breath of air. There is an irrevocable technological link between the atmosphere and breathing that lends itself to political and economic control [10].

The problem of oxygen is the problem of extra-terrestrial economics and, more generally, extra-terrestrial liberty [11]. If oxygen can be supplied within the framework of society that seeks to maximise freedom for its members without degenerating into a dictatorial state, then all other products can be supplied in the same way. Other products which require a much more developed infrastructure than we normally require on the Earth, including water, can be supplied under similar social arrangements and general economic precepts.

The resolution to the problem of oxygen must be the prerequisite for political freedom, quite beyond the problem of economic freedom. It is unlikely that any meaningful political freedom can be achieved in an extra-terrestrial society if the supply of oxygen is controlled by power-hungry coteries of individuals who use this commodity to exert exacting control over people and the state, or worse still, if the state itself is the fortress behind which these individuals use oxygen to solidify the control of the state over people and to coerce the population [12]. Surely no system of political freedom, even the most vigorous democracy, can survive in space if resources vital to the second-to-second survival of people is the preserve of tyrannical institutions and individuals? Therefore, besides being an economic problem, the central political challenge of a maximally free extra-terrestrial society is achieving the free supply of oxygen with a minimum despotic overburden.

Before condemning all extra-terrestrial settlers to a hopeless future under the control of the various organisations manipulating oxygen supply, it might be recognised that this terrible reality can be turned to an advantage. So extraordinary is the challenge of creating a free economic environment in which oxygen, and other commodities, are not used as levers of tyranny that a great incentive is offered to those in space to create and defend the institutions of the free economy and the political institutions alongside them. So alert and vigilant will they be to infringements and dangers to the free supply of these assets that they, of all people, should resolutely cherish and defend the institutions of liberty.

From their experience may well emerge more invigorated mechanisms to reduce the tyrannical oversight of economic activities. Equally importantly, they will have an intensity of focus on which to direct debate, discussion and argument that we usually lack on the Earth, except in times of war [13]. Oxygen will not only provide the breath of life to people themselves, but it will provide the political fuel for a combustion of argument concerning the construction and maintenance of the conditions for economical and political liberty in space that will permeate to every other consideration of the social system. These experiences may even beneficially inform the governments and social systems of the Earth, providing new insights into how to strengthen and improve mechanisms for maximising economic and political freedom.

The problem of oxygen is a classic conundrum of distributive justice, made acute by the extra-terrestrial environment. Any industry that seeks to produce oxygen must invest huge resources in people and infrastructure to extract this resource from rocks or the atmosphere. Once they have done this, then they will justifiably claim that they can sell and distribute this commodity in any way that they please. To expropriate this resource forcibly in the interests of the general public would, at least according to the libertarian view, be nothing short of theft – the theft of private property. On the face of it, then, such a view would seem to support individuals and groups that seek to set up oxygen-producing machinery and then to dominate that resource in any way they see fit. In such a situation the end point of a liberty-seeking agenda would in fact be to facilitate tyranny in a way that is less likely to occur with many other industries of less import to human survival.

There are three possible centrally controlled solutions to this problem. The first is indeed to expropriate this resource in the common interests and distribute it to the general public, but then what motive would any person have for setting up such an industry if they are constantly and unpredictably faced with the possibility of its apprehension by the state or some other body? What body of people would bother to place vast resources and effort into extracting oxygen when the rewards are its commandeering by another body and distribution to others? The second solution is to bring this activity under state control, but as I have elaborated, this is a recipe for totalitarianism; any single entity entirely controlling oxygen production and distribution will wield extraordinary power over people and this surely cannot be permitted, however well-intentioned the organisation may be initially. The third solution is some type of public ownership arrangement, a co-operative type organisation that oversees oxygen production and in which the public have share holdings (or their equivalent) in the company to wrestle some of the responsibility away from a central state body and into the hands of as large a number of people as possible. This third solution depends entirely on an active and wide public involvement in management of the organisation. It suffers from two weaknesses:

1) the possibility that the public will not show enough interest or exert enough control over it and it will degenerate into something no different than a state-controlled central entity,

2) it is still liable to become a consolidated central structure anyway.

Wherever, in any extra-terrestrial settlement, oxygen production and distribution is overseen by one organisation, then a tyrannical state of affairs is likely to emerge. The fourth option available is to give people the incentive to invest the resources in oxygen production and distribution, but to maximise the culture that leads to multiple entities producing this commodity so that no single group of people wields control over the possible sources of oxygen available. In this way competition will dilute political power in the industry, the possibilities for economic monopolies or oligopolies can be minimised and the likelihood that oxygen must be expropriated from anyone to supply the needs of the populace will be reduced.

It is worth pointing out that apart from the pursuit of a maximally free society, multiplication of oxygen producing and distributing entities has a very prosaic and quite pragmatic advantage that has no relation to political philosophy or economics – it will provide redundancy in production during an emergency or failure of a major producer, preventing catastrophic reliance on a single entity. Indeed, this crucial reality may itself be sufficient justification for finding the best social and economic mechanisms to maximise the number of organisations producing this resource from the very earliest stages of the establishment of extra-terrestrial outposts. It is this solution to oxygen production that I will explore here.

4. The Transition from a Centrally Planned to a Maximally Free Extra-Terrestrial Economy

In an extra-terrestrial environment, a maximally free economic system will not emerge from a large industrial infrastructure that existed before, changing hand-in-hand with a growing political and social enlightenment, as has been seen on the Earth since the seventeenth century. Rather, the economic infrastructure will, quite literally, land in an alien environment and become established. The important question is: what type of economy will this precursor economy be and hence what can we expect of it, or what should individuals try to do with it, to maximise the chances for liberty to flourish within it as it evolves and grows? We might aver that whether the initial infrastructure is established by corporations within an essentially free market or by governments, the initial state of affairs is likely to be similar – rare economic resources and consumables required for human survival will be owned and distributed by a few people or groups. It will in essence be oligarchic and centrally planned, even if the means of distribution is via a free price mechanism. The tendency for this initial state of economy to solidify into a central type of command economy derives from both the practical necessity of the environment and from human behaviour.

As supplies will be limited it will be necessary for an intense type of coordination. Imagine, for instance, a small colony on another planet which finds that it has run out of a vital and irreplaceable spare part for an oxygen supply system. It might muddle through using other space parts, but the danger that such situations present is far worse than on the Earth. The fear of such situations will impel people to set up, and universally agree to, a system of strong central control and accounting of all commodities to ensure that no single individual, or even society itself, goes short of vital supplies.

If the initial economy is one driven by private corporations for some economic benefit that they perceive (tourism, a natural resource that they have found, etc), then to protect their staff or the tourists under their charge they will be forced to implement a powerful system of resource accounting and documentation so that every stage of the process of establishing the settlement can be monitored and disaster averted. If the economy is one driven by a government initiative, then they will have a public responsibility to the people who have chosen (or been induced) to take part in settlement. The end result of either of these scenarios is the same in general character, if not in specific detail or practical mechanism – resources will be controlled and distributed to ensure a parity of availability to all people that require them.

A common criticism of the idea of a centrally planned socialist economy is that it is beyond the scope of any central authority or individual to carry out the vast number of calculations required to discern how to distribute and divide resources in society, even if they did have access to all the information required to assign values to objects, itself an unlikely proposition. Further, the rapidly changing conditions within society mean that no central authority can react rapidly enough to price and value fluctuations to maintain a viable form of fluid central control, a criticism that does not apply to the free economic price system, where each individual entrepreneur does respond, on a day-to-day basis, to changing conditions in the market [14].

Whether any of these arguments apply in economies on the scale observed on the Earth need not distract us too much here, but it does seem to be the case that central planning is practical in economies of a few hundred to a few thousand people [15], where the number of equations and calculations that must be carried out on a limited range of products is a tractable problem and can be adjusted in relatively short periods of time. It seems dangerous to hang an argument against central economic control on the supposition that these calculations are beyond the capability of the state when, in small societies with huge computing power and rapid telecommunications, it may well be possible for a state to both gather this data and respond to the changing economic equilibrium conditions within a society by modifications to production, distribution and pricing. It seems to me that whether the state can do these things or not is irrelevant. The important fact is that central control, successful or unsuccessful, hands control of the extra-terrestrial oxygen industry and other vital products to a small band of people who will have unfathomable powers to behave tyrannically. For this reason alone the central control of these industries should be rejected, whatever the technical capabilities for a rationalised and planned economic distribution of goods [16].

This conclusion takes on a particular resonance when the case of oxygen is considered. Although oxygen is one product likely to command totalitarian tendencies, unlike many other products the amount people require per day, and even per second, can be defined exactly. Unlike other products, the needs that people have are not defined by fashions, personal whims or desires; they are set by physiological constraints and people's changing needs can be quantified according to whether they are at rest or taking exercise. Given a certain number of people it is quite easy to quantify the amount of oxygen that they will collectively need. Now this may superficially support the idea that oxygen can be satisfactorily controlled by the state since even if all other industries are private, controlled and responding to a price mechanism to deal with the unpredictability of people's needs, oxygen needs can at least be prescribed accurately and can therefore be organised. However, this is an illusion. It is precisely this unfortunate duality of essentialness and quantifiability that will give oxygen production and distribution the tendency to be submerged into state control. The possibility for an exact regulation of oxygen production according to the rubric 'each according to his needs' is the very reason why the temptation to allow a highly controlled, centralised production should be resisted. Another argument for state control will emerge from the fact that oxygen, unlike water and food, cannot be easily divided between people and so the costs of use are difficult to apportion to individuals. This logistical fact may encourage authorities to argue that private production is unfeasible and that it should be under common ownership.

There is a sociological reason for resisting central planning.

The encroachment of state power into extra-terrestrial industries that lend themselves to centralisation is apt to encourage a type of ennui and resignation in the population. Once an industry as important and pervasive as oxygen production is controlled centrally there is less impetus for the population to challenge it and there is more likelihood that they will accede to the control of other vital industries and retreat into apathy with respect to activities that are in conflict with the state. The same arguments adduced by the state for the control of oxygen will be applied to other industries. Thus not only does the smothering of one industry in a blanket of state regulations make it easier for the state to use these same mechanisms to influence an ever-growing sphere of industrial activity, but the people themselves become used to this mode of operation and then accept growing state control as the way things are normally achieved. A type of positive feedback is liable to set into an extra-terrestrial settlement, itself underpinned by the persuasion that the state will be the greatest guarantor of safety in these industries [17].

Now the reader might properly point out that a particularly enthusiastic and militant group of entrepreneurs with no regard for safety and a quite laissez-faire approach to space exploration could successfully establish a large settlement where everything was free for the taking, or at least open to commercial transaction, and there was very little central control. Although this situation seems highly implausible, the most negative starting point is taken here in order to achieve an analysis of how a free economy might be constructed from starting conditions that are essentially centrally planned. Any lesser form of control and a starting situation of greater freedom is, it can be assumed, a good thing. The question is how the early centrally planned economy is to make the transition into a large settlement where economic liberty is maximised.

The difficulty in achieving this transition is made acute by the inertia that is characteristic of centrally planned societies. In the case of central planning governed by a corporation, there are profit motives for retaining control of resources partly because control of resources either directly implies control over profitable assets or because resources, such as habitats in extra-terrestrial settlements, may be used as the source of profits, for example through tourists. A secondary inducement to control is the fear of losing future profits which may come from existing assets, but are currently unpredictable or unknown. The greater a sphere of activities and assets that a corporation presides over, the fewer opportunities for future profits will be lost to others who gain access to those resources.

Added to these deliberate systems of inertia is the practical problem of the adjustment in systems of production and distribution. A corporation that produces and distributes large quantities of materials and products for life support systems has in place a reliable and interconnected system of accounting and organisation in a lethal environment with which few will be willing to interfere. The situation for a government-run settlement will surely be much the same. The inducements and inertia within the system will ineluctably result in the same level of control and reluctance to abnegate any part of the system which might ultimately threaten safety. In most cases the trade-off between the potential disruption of the economic system and sustaining the *status quo* will almost invariably fall on the side of the *status quo*.

These are the conditions from which a free extra-terrestrial economy must then emerge. How can the scions of the initial extra-terrestrial society achieve this? The problem can be more simply approached by considering again the specific case of oxygen. No matter whether the supply of oxygen is initially by government or private industry, the priority must be to wrestle control from a single authority and maximise the multiple sources from which this commodity can be derived, thus reducing the power that any single individual or organisation has to use oxygen as a source of power [18]. The simple solution would be to maximise the conditions for competition by creating a free and unhindered price mechanism, whereby new players in the market can undercut others and produce oxygen more efficiently or at lower cost than existing producers. The first stage to the reduction of the chances of monopolistic control can simply be stated as the creation of the political and constitutional mechanisms to protect a free capitalist economy in which any individual or group can set up oxygen-supplying ventures and challenge organisations in the field [19].

From the earliest stages of the extra-terrestrial society an environment where people can gather, establish new enterprises and compete against existing enterprises must be encouraged, and this activity must be enshrined within the political doctrine and constitutions established. It will require a foresight of a greater social good for organisations to relinquish monopoly power in the interests of creating a free economic system. The original organisations involved in establishing an early extra-terrestrial settlement must submit to this system, or, even better, take part in the creation of the system from the very outset of the formation of an extra-terrestrial settlement. It is a matter for jurists and economists to frame a system that will establish the formal legislative mechanisms that allow competition.

However, as in the case of monopolies on the Earth, there is still no guarantee that the systems of supply and demand would not, at some stage, be gathered together by a single organisation either intentionally by takeover bids, or by a slow accretion of resources through practical expediency. The problem with oxygen is that no latitude for such an eventuality can be allowed; it will inevitably lead to a lasting loss of economic freedom, the encroachment of tyranny and a dire threat to political freedom from which society will find it difficult to extricate itself [20]. Monopolies commissions and other such bodies guard against this on the Earth and these organisations provide some model of how this might be achieved in extra-terrestrial environments, but it seems that a more energetic approach to this problem should be implemented in space.

To vigorously develop the free economy, the extra-terrestrial state might oversee a mechanism of competitive fragmentation achieved by breaking up large organisations whose control has become so extensive as to warrant concern [21]. It might also provide tax and funding incentives to new enterprises established in 'critical' industries. The purpose of such initiatives would be to maximise competition and reduce the chances for monopoly. Again, these activities have their parallels to state-forced monopoly break-ups on Earth, but the identification of critical areas of industrial activity in which a special effort was provided toward new enterprises would be an additional procedure. These critical industries would include: oxygen production, distribution and maintenance of oxygen-producing machines; food production and distribution; water production and distribution; machine maintenance; space suit production and maintenance; and habitat building and maintenance.

The state or its terrestrial corporate or state precursors might go further and provide the very infrastructure and equipment to individuals and groups that wish to go into production. As gathering together these practical requirements in any environment in outer space is logistically prohibitive and easily usurped by large organisations that already have access to these resources and the means to control them, the state could facilitate the provision of these capabilities to competing entities (perhaps by state loan agreements or facilitating private venture capital), encouraging, from an early stage, plurality in the means of supply.

These industries could be required to adhere to principles of complete transparency. Accounts and major transactions could be publicly available information and members of company boards and senior officials could be required to present publicly any conflict of interest or involvement in other industries of a similar nature. This mechanism cannot prevent collusion between industries and the formation of cartels. A maximally free political society cannot monitor or control the ability of corporation heads to meet and talk about any matter they choose, but transparency can at least maximise the chances of bringing to light any nefarious deals aimed at excluding competition and monopolising oxygen production. This approach amounts to a philosophy of developing the maximum equality of opportunity for members of an extra-terrestrial society to gain access to information.

A board could be established, run and overseen by elected members of the public who would oversee the oxygen industry and its activities. It would have the power to request information on transactions from industries and the state alike and to seek enquiries on any aspects of the industry it found to be of concern. To prevent it becoming a weak and inferior organ it should have the power, through the judiciary, to request prosecution, and within it would be a limited number of state officials through which this strength would be possible. A natural question is: how are we to ensure that this board itself is not usurped by the industries and the state? In practice the answer cannot be anything other than that we cannot guarantee this – other than providing the specious answer that yet another board should oversee this one, *ad infinitum*. Such a policy would risk the multiplication of organisations, so beloved of tyrannies [22]. As with the separation of powers at the political level, the system can never guard completely against corruption, but it can provide for a greater fluidity of mechanisms for oversight and maximise the chances that individuals and organisations with concerns have a mechanism to speak.

Beyond these considerations, it will be mandatory that people in extra-terrestrial settlements live in a culture of wariness towards the oxygen industry and other vital enterprises. They will need to take an active and vocal interest in the affairs of their industries. One purpose of the state, which I will return to later, is therefore to create the conditions that maximise the active involvement of the public in the decision making processes of the wider polity [23].

Beyond the oxygen-producing industry, but not disconnected from it, are similar concerns with the health and safety sector. The reader might point out that the tendency for health and safety to become controlled by monopolies should make it more of a concern than the oxygen industry, since it will be applied in so many industries from spacesuit production to habitat construction; the oxygen industry may constitute only part of its reach. In the case that some sort of overarching state health and safety executive be created that oversaw all production, a possibility which should be vigorously guarded against, then this argument would be true and the same concerns raised about oxygen would even more forcefully apply to such an organisation. However, even if the state should successfully fragment health and safety regulations into their respective industries, the concern that they will become a means for control over those industries in the name of protecting individuals and society as a whole to a much greater extent than on the Earth.

The same system elaborated for the oxygen industry could be applied to health and safety industries and organisations. They could be private firms, disconnected from the state and open to competition. Health and safety is one of the most profound threats to liberty because it can be used both as an instrument of positive liberty by the state, but at the same time as a means to coerce people into good behaviour by the false pretext of protecting them against the impending failure of oxygen systems and other life support systems. The bureaucracy that will tend to emerge from the requirement for health and safety systems and the attraction that this requirement has for propagating a network of officials means that health and safety should be removed from state control to the maximum extent possible [24], except for the setting of established standards in service provision and equipment.

To prevent health and safety officials from assuming vast powers and expanding the realm and complication of edicts, health and safety could be offered by corporations that compete under the state-guided quality standards. Each corporation would offer packages of health and safety oversight to individuals and corporations that they could choose between. Some of these offers may be substandard and it is conceivable that lives might even be threatened, but these corporations would soon go out of business. They will not attract new customers by asphyxiating past customers.

These proposals are not ones that are purely a theoretical consideration of the direction that an extra-terrestrial society might take in the long term; they have useful practical implications that can be considered from the earliest stages of the building of an extra-terrestrial settlement, even the first outposts. The architectural ramification of the preceding discussions is that from the beginnings of an extra-terrestrial outpost habitats and other infrastructure should be modularised so that no system of health and safety can smother large numbers of individuals. Modularisation of the infrastructure is the best hope for the emergence of a system that allows individuals choice. The modularisation of the physical infrastructure makes it easier to fragment food production, water preparation and recycling, and oxygen delivery; in this way it becomes more possible for 'districts' to become established and local corporate entities to offer services to different habitats, maximising the diversity of economic activity and reducing central oversight. If oxygen supply, habitat services and other means of distribution are all centralised physically, then it will be difficult to achieve open competition; society will naturally tend towards a central form of organisation simply because completely interlinked logistics and resource allocation lend themselves to a central auditing, accounting and supply system. This matter provides a particularly clear illustration of the reason for debating the nature of extra-terrestrial freedom even prior to the establishment of settlements, because the expression of liberty may well be maximised by the incorporation of certain characteristics into the very architecture of settlements [25]. Thus, we can argue that the promulgation of liberty should be a guiding objective of individuals in space today and during emerging human space exploration efforts in coming decades, in preparation for the subsequent strengthening of liberty in future centuries.

All of these proposals may seem like those of a fully developed nation-state, but there seems to be no logical reason why they could not come into being in the smallest extra-terrestrial society. In the first stages of society the lack of large numbers of people will lend itself to corruption by dint of the fact that the same individuals or their close relatives and friends would have involvement in the different mechanisms of supply and distribution. As society grows larger, competition in the means of supply and plurality in ideas about how to do this becomes more easily realised. One means by which a matured society can be directed towards this end and away from a centrally controlled system is by a system of education.

5. Education

The environment of outer space and planetary bodies is not one that is conducive to the development of an expansive vision of society amongst people. Their vision of the world will be circumscribed by a narrow range of experiences within habitats and pressurised modules, apart from rare opportunities in spacecraft and rovers or spacesuits to visit some equally barren region in space or on the surface of a planetary body. The lack of any widespread biota on other planets will limit the experience they have of biological diversity and thus the richness of their view of the phenomenon of life.

Coupled to these restrictions on outlook will those imposed by the social and political requirements. These influences will find their greatest effect on young people, particularly caused by the pressure to maintain youth as good and responsible citizens. This compulsion will not necessarily come from the edicts of a despotic state, but from the desire of their elders to ensure that they have not reared individuals who are errant and a threat to others in a dangerous environment, itself underpinned by powerful peer pressure. It will be a long time until young people are born and brought up in space, but this challenge also applies to adults and their continuing education.

From these influences and inducements comes the task of creating free-thinking individuals who are willing to challenge the *status quo* and pursue new intellectual and practical paths for themselves and others. How is the extra-terrestrial state to be prevented from seizing upon education as a tool to assert its authority over the extra-terrestrial society [26] from the education of young people to the education of adults in new vocational skills and higher education?

There are a range of educational requirements which cannot be devolved easily from state oversight. The technical complexity and artificiality of extra-terrestrial settlements make it mandatory that individuals, even more so than on the Earth, receive a basic training in the sciences. Their oxygen and water will be the product of artificial physical, chemical or biological processes, their food will be the product of biologically controlled processes. Whereas the latter is also the case on the Earth, the extent of dependence on these systems in outer space will require a society in which individuals at least have some rudimentary understanding of how these systems work and the fundamental scientific and technical principles behind them.

But more than this, the education of people in the scientific skills that underpin the functioning of an extra-terrestrial settlement will be the very mechanism by which liberty can be secured, because with this knowledge they will be in a position to challenge claims about the necessity of health and safety mechanisms and state oversight in various industries. They will also be in a position to suggest, or even demand, new technical systems and processes that will improve their life support systems and general well-being. With a population that lacks detailed scientific and technical knowledge of specific industries, the state and all its officials are left with the opportunity to wield almost wizard-like powers over a people who stand ignorant and in awe of the vast industrial complex that sustains their lives [27]. More so than on the Earth, in extra-terrestrial settlements scientific knowledge will be the educational guarantee of liberty.

Provided that scientific knowledge and information is not politicized (and this is not necessarily a given condition, as arguments about creationism and evolution on the Earth attest), there is no reason why the state could not provide education in much the same way as state schools and higher and vocational educational establishments do on the Earth. The standardisation of this education, in terms of curricula and activities, can be accomplished by some form of central coordination.

So too with artistic education. On the Earth, even the most culturally illiterate person is exposed to enormous artistic diversity, both manifest in the biota of our planet and its wonderful and fantastic forms, and the exposure to ideas and opinions that are the combined melting pot of tens of thousands of years of intellectual advances, military conquest, enlightenment, strife, natural disasters and the birth and decay of civilizations.

The lack of prior history and social order in space and the utter monotony of the landscape will simply eradicate the natural artistic and cultural education that people are inclined to benefit from on the Earth. Art and culture must be a pervasive part of education in the extra-terrestrial society, offered not just to the youngest members of a society, but to all persons. Its purpose must be to overwhelm the historical and environmental emptiness of the extra-terrestrial society and to fill the void with ideas and an intellectual eagerness–an intellectual onslaught against the potentially crushing conformity of outer space [28].

In this view I share a common approach to political philosophers who have supported the idea of 'state perfectionism' in which the state's role goes beyond the task of merely upholding justice and other basic tenets of a liberty-seeking society, but in the extra-terrestrial environment it is justified in supporting and subsidizing artistic and scientific ventures that seek to promote cultural and technical education as an antidote to the tyranny that might be caused by the natural environment. However, the state must support these activities in the most general way possible and it should not pick and choose activities that will amount to a state-sponsored education agenda advancing specific ideas and special interests 29.

The space environment will have little political and social history. Although terrestrial history may seem remote to people, it is the only discussion available to a society that lacks its own history, at least in the early stages. To ignore this history would be to consign individuals to a type of pre-Enlightenment ignorance of where the concepts of society, including liberty itself, have originated and the enormous sacrifices that have been made to realise their own existence. Teaching people terrestrial history would not be an imperialist imposition from the Earth, but instead a factual basis for them to understand from where the exploration and settlement of space emerged and how the social and philosophical concepts that carried them to this point might be melded and changed to suit their own society.

The serious problem that the extra-terrestrial environment represents to free-thinking individuals, particularly young people, can be addressed by augmenting a basic educational curriculum with a diversity of schemes, some of which are familiar on the Earth. These may include the opportunity for private education provided by people with specific interests and skills, thus enriching the opportunity that individuals have to gain new skills and perspectives independently of the state. Private education may be one effective way in which individuals can be exposed to the many ideas and views that will mitigate the natural tendency of the extra-terrestrial environment to limit them mentally. A valuable means to diversify experience will be to provide central funding, perhaps in the form of vouchers, that can, in the case of young people, be redeemed at institutions for their education [30].

Administratively, educational establishments and their curricula must be overseen by elected non-state officials to minimise the chances of state manipulation of curricula and ideas. Independent boards that would both devise and oversee curricula could be created. The role of the state would be reduced to helping to define technical subjects and other information that is necessary for individuals to understand the workings of an extra-terrestrial society at the scientific level. Any individuals could stand for election to these boards and the state would have no power to prevent the appointment of individuals with ideas counter to those of the state. The prevention of quackery and charlatanism within the system would be achieved by a faith in the good sense of all people, particularly parents, to believe in the value of objective education. This would be sufficient to ensure a generally robust electoral system.

It is impossible for a person, when born and brought up in a particular environment or political system, to objectively assess the extent of their liberty, unless they have some clear comparison [31]; this is a 'cryptic natural tyranny' [32]. A purpose of an extra-terrestrial education system is to expand the vision of individuals relentlessly and so to guard against the assumption, which may emerge at any time, that maximum freedom has been achieved and liberty successfully given its greatest expression. If one assumes that the liberty one now experiences is hidden under a veil of tyranny which is invisible precisely because one is living within it and therefore ignorant of it, then there can never be complacency; society is striving to discover unknown tiers of ignorance and remove them. The recognition of this reality must be one of the key underpinnings of the educational system of extra-terrestrial societies [33].

The objectives of an extra-terrestrial education system can no doubt occupy the minds of educational planners to a much greater extent and in more detail than can be done justice in a short essay, but it may be sufficient to summarise by noting that whatever those details are, including the suggestions made above, they are all underpinned by two principles that should guide any initiatives established in support of extra-terrestrial education – first, to minimise the chance of state subversion of the educational system to control the population and second, to maximise the extent to which individuals can be emancipated from the natural tendency of the extra-terrestrial environment to create restrictive and tyranny-prone views of society. The latter problem amounts to the problem of the tyranny of the majority – the levelling of the population by the environment and culture it spawns into an amorphous mass of conforming entities brought under the control of common laws designed to protect the individual, but in fact enslaving it [34]. The role of the state and the education boards in education is to stand between conformity and educational anarchy by ensuring a basic common level of knowledge and expertise necessary for individuals to survive in the extra-terrestrial environment, but to ensure a sufficient incentive toward diversity of views and opinions to prevent a majority from holding sway over the non-conformist. This may sound no different from the objectives of a terrestrial education system, and there is no claim here to the contrary – or that in fact this is even an original view – but the extra-terrestrial environment does create a unique and powerful force in the direction of dictatorship that must be halted by vigorous efforts through the education system that have previously been unseen on the Earth.

6. The Culture of Liberty in Society

In many of the considerations above there has been an apparently weak inference that preventing tyranny requires a culture of freedom amongst people. This is not some overly optimistic and simplistic utopian hope. There are ways in which an environment can be created that eschews unnecessary hierarchy and maintains linear management structures only to the extent that expediency requires them to carry out vital and demonstrably useful social, political or economic functions. The history of the tyrannical use of arbitrary power structures is rich, particularly during the medieval period, when primogeniture, inegalitarian taxes, the guild system and countless structures and interest groups supported by the monarch, the aristocracy, and innumerable other combinations of vested interests wielded unrelenting power over the vast masses of people in Britain and the Continent.

The triumph of the pursuit of liberty beginning in the latter part of the seventeenth century was the removal of these structures to the maximum extent possible and the development of an environment in which individuals could express their creative and industrial powers, even if the path was not completely without tragedy. This transition was helped by the unintentional influence of the Industrial Revolution and the affect it had on reducing, in fact abolishing, the power of the guilds. Given the propensity that extra-terrestrial societies will have to err toward tyranny on account of environmental conditions, the continuation of this general philosophical direction is the best insurance that future settlers will have against despotic power structures.

The nurturing of a culture of liberty in the extra-terrestrial state could be accomplished by an ethos embedded within law and encouraged, through a constitution, within the population itself, which rejects power and inequalities in society that are not absolutely necessary and useful to the achievement of a social purpose, and whose value can be shown to be beneficial to the common weal. This latter requirement is essential as any type of power can be defended as potentially useful to someone or some special interest group. What allows one to separate a power that is unjust, such as primogeniture, from power that is not, is the characteristic that distinguishes power designed to channel influence in one special interest direction from power whose objective, and demonstrable outcome, is to better the lot of all. When judged against this metric, the extra-terrestrial society, as on the Earth, will fare better in creating an egalitarian society, where by egalitarian I do not mean economic equality, but equal under the law.

The creation of an egalitarian society that eliminates gratuitous privileges, hereditary control and arbitrary power seems capable of making enormous strides towards providing the culture for extra-terrestrial liberty. In addition to the more general culture to be created, there are some specific institutional objectives that can be identified, of which one, which should be woven deeply into the structure and legal systems of an extra-terrestrial state, is touched on here.

A notable influence on the progress, and the means to progress, within a society, and one which is no minor matter, is the way in which society recognises achievement. Any person with reasonable powers of observation can see the affect that reputation can have on the influence that an individual has in society. Extraordinarily good and intelligent people are sometimes not well known. Sometimes, rather mediocre people seem to wield great influence and weight in society. This curious reality derives from the tendency that people have to be awed by influence and the acceptance that people seem to accrue, particularly as a result of their deliberate effort in this direction (which is a reason why mediocre people often achieve power, because they spend their time nurturing acceptance from authority, which intelligent, independently minded people are rather more inclined to ignore and maybe even actively shun [35]). This phenomenon has great appeal to the state and its officials, and many non-state institutions, as a mechanism for social control. By conferring honours on people whom they regard to be useful, they create a self-perpetuating system of support, since honours will make these people more highly regarded by the less thinking and critically analytical elements in the rest of society and therefore in turn more useful to the very people or organisations to whom they were considered useful in the first place.

A system of recognition by the state therefore feeds on itself. Individuals seek to gain honours to be part of the system that decides who should be recognised and others seek to gain honours to be recognised by those that confer these honours. Carrying out good deeds and achievements is no longer a satisfactory end in itself, but rather is additionally the means to achieve recognition. In other words, the edifice of social success and recognition becomes arrogated by the state. Now whether fair-minded people ever actually manage to create a state-sponsored system of honours and recognition that seeks no power for itself, but merely the genuine recognition of good in society, is beside the point. The matter under deliberation is whether, on the matter of principle, it is right for a state to even intervene in the matter of recognising who has done good things for society, even it is does believe it can do this fairly.

In a politically and economically maximally free society many activities are against the state; this is the very basis of a society that pursues the tenets of freedom. Indeed, the progress of a maximally free society is constructed on the idea that individuals or groups of people can propose heterodox ideas or positions that do not conform to the state's view and they can be recognised by their fellows. This establishes what we might regard as a basic principle – that the state should seek no part in recognising the achievements of individuals since by definition those achievements are not supposed to be aligned to state interests, but only so aligned by chance because the actions and opinions of an individual happen at one particular point in time and place to be the same as those advanced by the state.

The danger to freedom lies in the problem that a state-sponsored system of recognition encourages people to do things that achieve the approbation of the established institutions and thereby it distorts and discourages the propensity toward free thinking in the population. In its turn, the system encourages the state to use its rewards system to subtly display disapprobation towards those individuals it considers to be a threat or to have done things of which it does not approve; and it encourages individuals within the system to show a similar disapproval to demonstrate their credentials as reliable and stalwart members of the system. Under extra-terrestrial conditions these mechanisms become a serious threat to liberty.

How, then, is anyone to be recognised in an extra-terrestrial society? Assuming that people need some formal recognition, the best means by which this can be achieved is by an organic process, whereby success in any given field or activity is derived from peers and members of society who appreciate whatever accomplishments we are talking about. One would prefer a society in which no one even needed recognition, but where instead people were content to do things out of a plain and simple enjoyment in expressing their own abilities and interests toward the benefit of social progress. However, it must be accepted that some form of feedback and affirmation from society is regarded as desirable by most people.

An organic process of recognition already occurs in society, particularly in academic fields where one's reputation is derived from original work published and read or admired by others. The advantage of this organic process over a state-sponsored process is that it generally, but not always, obviates the possibility of sycophancy to some central authority or body of people to increase one's chances of receiving an award. The more recognition and reputation are left to a natural process, the less liable are heterodox ideas to fall prey, early in their emergence, to state-sponsored disapprobation.

In an extra-terrestrial society the resolution to these complications is to abandon forms of state recognition, thus removing from the state a powerful means to consolidate an established order and a weapon with which to manipulate public views of particular people, either for good or bad. In this way the extra-terrestrial state may emancipate itself, and the people whom it serves, from hierarchies which will otherwise provide an irresistible conduit through which state officials and others may seek to determine who is influential in society.

7. Religious Institutions

The extra-terrestrial environment is so lethal and barren, regardless of where one chooses, that it is the most fertile ground for the seeds of religious doctrine to take root. The view one has of the existence of a god does not change the fact, amply demonstrated by history, that religion has been a vital and guiding force in human affairs and individual lives for millennia. Apart from people who envisage a world freed of religious belief and its arbitrary testament, it is probably much more sensible to realise the value that religion has in human lives and, accepting this to be the case, to find a way to canalize it in the most positive way in the extra-terrestrial environment.

There can be little doubt in saying that the most grave threat to extra-terrestrial liberty that may come from religion is the imposition of an all-embracing religious doctrine on an entire community; in essence, a theocratic state [36]. If the reader should wonder why this apparently negative and extreme possibility should be a matter of immediate concern it is worth simply considering the environment under which extra-terrestrial societies will emerge. Their populations will be small, they will inhabit extreme environments that threaten instant lethality on a day-to-day basis and the barrenness of their environments and the confines into which they will be forced will be spiritually draining. They will seek a spiritual release.

For a feeling of camaraderie and common bond to be established, people will crave a common spiritual release, since worship on one's own is lonely. It is easy to see how a religious doctrine, held by all, would become established within a settlement in its early stages and permeate and propagate to later stages of a settlement as it expands and grows. A conformist and powerful religious doctrine that amounts to a theocratic extra-terrestrial state that exerts enormous and uncompromising control over people is a dangerous and likely path [37]. All attempts must be made from the outset to prevent this; once it becomes established and buried within the fundamental view that people have of their place in the extra-terrestrial environment and the Universe at large it may take decades, even centuries, to displace [38]. There is also a wealth of evidence that many people regard religious doctrine on the Earth as responsible for liberty. Although this is a defensible position, in the sense that history provides good evidence that religious belief has played a part in the development of modern Western cultural attitudes [39], it is easy to become so convinced of the formative role of religion in liberty [40] that one encourages it and then defends it on this basis, such that those who oppose a particular religious doctrine are then perceived to be against liberty.

So the extra-terrestrial society must embrace religion to allow people to express a need for worship, but it must hinder and stop widespread religious belief that takes advantage of the extreme environment of space to perpetuate tyranny.

In much the same way that has been achieved in some states on the Earth, the state, religion and education must be separated. Some mechanisms by which education would be partially separated from the state have already been discussed. Religion should also be treated as a pastime, exercised by individuals as a private concern [41]. No type of curricula should include forms of religious doctrine and no religious groups should be permitted to set-up educational establishments. Educational groups may be permitted to describe religious beliefs in curricula, but not to advocate them. In other words the state should play no role in advancing religious beliefs.

But the extra-terrestrial society could go further than on the Earth because of the special threat that the extra-terrestrial environment creates in the expression of religious belief. As with education, independent boards could be established that seek to identify and remove religiously inspired dictates and policies from both state policy and educational curricula. Individuals may be elected to these boards, which could establish a maximum percentage representation of any religion, should people with religious affiliations choose to join them. This may prevent these boards from becoming usurped by religious groups.

These boards themselves would not be anti-religious or tyrannically inclined to hinder religious belief and they should not become instruments for witch hunts against the development of religious thought, but they could act with objectivity and sensitivity to remove religious doctrine from the non-religious organisations of society. They could act to push extra-terrestrial society away from religious incursions that may threaten to make state policy or education a servant to religion. In this way, society could incorporate religion, but the greatest religious challenge to a plural and maximally free society could be vitiated from the earliest stages of the formation of extra-terrestrial settlements.

8. Private Land and Property

The commandeering of private property and land becomes an enticement much more real to those who lack that property in an environment where the conditions are lethal. However, the usurpation of property and land by the state can no more be tolerated in an extra-terrestrial environment than on the Earth if the conditions for liberty are to be nurtured. Thus a paradox faces the extra-terrestrial state which revolves around the question of when it is legitimate to seize property in the interests of livelihoods and when it is not [42].

It is worth stating again the famous words of William Pitt, 'necessity is the plea for every infringement of human freedom' [43]. The simple answer to the problem of property ownership is that it should rarely be right for the state to expropriate property since it must set itself the prior objective of resolving any situation that threatens the lives of people before it resorts to seizing property [44].

One cannot imagine war in space, at least in the foreseeable future, although the long-term future may hold a different story. The most common excuse for the expropriation of private property is therefore absent and the only situation in which such extreme measures can be envisaged is when a body of people are threatened with death in the extra-terrestrial environment and no recourse other than the acquisition of others' property can resolve the situation. An imaginary debacle involving a failed oxygen supply system in which segments of oxygen systems, owned privately by others, must be commandeered to avert an impending disaster in another segment of habitats, might be envisaged. But as with the systematic state claim to private property on the Earth, such situations must be overseen by deep political discussion and misgiving.

The lethal conditions in space do not prevent corporations and other suppliers of commodities from selling their wares to individuals in a traditional type of transaction whereby the property is then within the private domain of the buyer; and a great deal of suspicion and analysis should attend the excuse that the lethal conditions require public ownership of this and that commodity. In situations where early public ownership of resources seems likely, such as food production systems, the state, and the people, must seize upon all means possible to expand the rate of production and accessibility of the resource so that corporate or private ownership and competition becomes possible. Thus, the same principle that applies to the oxygen industry discussed earlier underpins the very notions of private property in space.

We cannot apply exactly the same view to land, where some type of public ownership may be desirable in certain spaces [45], but apart from these rare preserves, land could be available for private transaction both so that private industries can acquire natural resources and use them to better the economic field of play of the extra-terrestrial settlements; and so that individuals can themselves seek space and resources independently of others. The transactions by which this land is traded may be by the same regulations as in many nations on the Earth: sold and bought by developers at prices that appeal to those who believe that they can do something useful with the land.

A more obvious question pertains to land that is not already under ownership – how is extra-terrestrial land to be claimed in the first place? This is a matter that has occupied a great many people, and despite all the complexities and arguments about planetary protection, UN legalities and so on and so forth, a simple Lockean response to this question [46] seems the most practical way to deal with the problem – any group or individual who can find a use for land and do something productive with it should be able to claim it. Only this policy will maximise the chances that individuals and corporations will risk themselves in the lethal conditions of space to create enterprises beneficially using the land and expanding the human presence in space. The legal condition would be that these people must do something productive with the land, rather than merely claim it in absentia, hence preventing vast tracts of land sitting unused, but claimed by absentee landlords waving pieces of paper with descriptions of their landholdings, a problem that has already manifested itself in the public sale of land throughout the Solar System as gift items.

In this scheme of land, the role of the state would be to collate claims on land and verify that productive activity is occurring on it, and to arbitrate and set the general laws that would apply to all subsequent transactions of land, which might be accomplished by estate agents or realtors no differently to the processes that occur on the Earth. On Earth today, this vision of extra-terrestrial land is rejected by some people, but the reason seems to be a product of envy – a distaste that land will be claimed by rich corporations or nations that have the spacefaring capacity and denied to those that do not. But what is the point of denying land to those who could do something useful with it to placate those who would prefer to see nothing done with it at all, simply to satisfy their desire to level the whole of humanity to the lowest economic denominator? Land in outer space, barring that of potential biological interest and placed within the remit of planetary protection concerns, and accepting some minor regulations regarding pollution and waste in all other lands, should be free for any person or group to exploit and develop. The control of land by international or national regulations to hinder its free use – and the restriction in its use under the spurious claim that it is the common heritage of all humankind – is a form of tyranny [47]. There is no meaningful argument that can sustain the claim that land is the province or heritage of all humankind. Indeed, from a general point of view there seems to be considerably more greed and hubris bound up in the idea that everything in the Universe belongs to humankind compared with the somewhat more modest claim that a few patches of land here and there, an infinitesimally small percentage of the material Universe, should be able to be claimed by corporations or individuals from planet Earth who can find something useful to do with it.

9. The Limits and Purpose of the Extra-Terrestrial State

From these previous considerations emerges a view of what the extra-terrestrial state is. It is a crucial task to define specific liberties of, for example, land ownership, the right to association, or religious freedom, but all of these liberties are only protected *en masse* if they are bound together in a philosophy that permeates society and defines the architecture of the state. Otherwise, individually, they become subject to decay or mistreatment. Here it is my intention to synthesise the previous discussions into a more general embracing view of the state.

The foregoing discussions show that it is the physical environment that overwhelmingly defines the behaviour of people in outer space and is most likely to provide the conditions and excuses for the emergence and maturation of despotism in society, from the birth of individuals, through to their education, religious expression and vocations.

The fundamental purpose, therefore, of the extra-terrestrial state is surely to interpose itself between the people and the extra-terrestrial environment. Its purpose for existence is to arrest the drift of the population towards vapid conformity caused by the extreme characteristics and monotony of the environment and the culture that health and safety and caution will tend to generate. It must seek to prevent the social contract forced upon individuals by the interdependent links in society required for survival from consolidating itself into an instrument of all embracing tyranny [48].

The challenge that faces the extra-terrestrial state and any of its precursors on a day-to-day basis is to nurture the advantage that isolation in an extreme environment can provide in generating a sense of community, but to encourage autonomy and individualism within this community that will be stifled by the natural environment responsible for the very generation of this sense of community. The perpetual and indefeasible challenge of the state is to vivify the population and the sense of individualism, whilst itself resisting the temptation to succumb and become a servant to the extra-terrestrial environment in order to serve special interests and itself. This, stated most simply, is the purpose of the extra-terrestrial state and the general culture and doctrine which it must pursue through its officials [49].

In the historical context the challenge of the extra-terrestrial state is to balance individualism with collectivism in the specific conditions of the environment of outer space. This core challenge is no different from any state that has existed on the Earth. This inherent social conflict in purposes was a challenge of almost insuperable proportions for the French between the period of the Revolution and the Third Republic and the drift and counter-drift from tyranny to anarchy that French society encountered. It is not clear that the balance will be any easier to achieve in outer space. Unlike early European societies, the challenges faced by the extra-terrestrial state are simpler. The populations, at least to begin with, will be smaller, the history will be less rich and the focus on a single parameter – the tyranny of the environment – offers a less complicated problem than the long histories of European nations.

The most predictable emergent vision of the state which will be inimical to freedom is the development of a Rousseau-like 'general will' within the population [50]. The environmental extremes are not peculiar to this and that individual, but are rather identical for everyone in any particular environment in space [51]. The development of common approaches to addressing this challenge, physically and therefore socially, is liable to encourage the slow consolidation of a socially cohesive will directed against the environmental extremes. As all individuals must agree that survival against the extra-terrestrial environment is not only vital for individuals, but also for the continuity of the society and therefore the state, it follows that the will of the individual is indivisible from the will of the state; both seek the same end [52]. It is certainly the case that in any society the people must be one and the same as the state, or have a strong sense of ownership of the state; otherwise the body of officials that constitute the state becomes a separate entity, detached from the people and more likely to regard itself as superior to them and above democratic process. But at the same time, if individuals become subsumed into a type of general will and entirely inseparable from a general credo enunciated by the state, then extra-terrestrial individualism will be destroyed [53].

Through diverse institutions, born through education and the culture of society, the general will may be eroded and the state may ensure a healthy independence of mind in individuals, through which a common response to the environmental conditions may still be achieved within the context of a socially and politically diverse state [54]. To examine this problem in more depth it is useful to identify two of the mechanisms by which a general will is likely to assert itself in an extra-terrestrial society.

The first concerns the tendency of states to bureaucratise. To minimise the tendency toward the imposition of a common will, the state must resist the temptation to develop excessive bureaucracy which will generate a technical state in which all details are carefully controlled and coordinated, stifling the emergence of a vibrant political state [55]. As an extra-terrestrial society is founded on technological capability, particularly in the provision of basic commodities such as oxygen, then it will have a tendency from the moment it is formed to substitute flexibility with an exact and defined process for each and every action that it undertakes. Clear protocols in safety and industrial manufacture reduce the chances of disaster caused by technical error and decrease the chances of crises caused by the ineffectual or distorted transmission of information through the industrial and political structures of society. These errors are most easily averted by ensuring that as many decisions and actions pass through as many people as possible to increase the chances that they are detected before they cause a disaster. Thus the drift of the state towards a focus on technical activity will also encourage an increasingly powerful bureaucracy to oversee this activity in the many areas of human industry that it chooses to oversee.

But it seems that the risks of devolving administration with its potential inefficiencies are the price that must be paid to achieve industrial competition and a diversity of political institutions that are the best guard against the sequacious conformity of a general will. Multiple corporations dealing in oxygen-producing machinery increase the likelihood of technical incompatibility and error, but their effect is also to multiply the possible solutions to the diverse health, safety and technical challenges in oxygen production. This in itself will redound to society by increasing the plurality of industrial and political debate concerning the development and progress of the oxygen industry and diluting the tendency of a convergence of opinions and practical solutions to oxygen production that will lead to a general will. It is within the technical realms of society that proliferation of organisational diversity would be a guiding objective of the state, both within private and public enterprises. As extra-terrestrial society will be faced by many problems, it seems counterintuitive, but quite likely to be realised in practice, that state control and the conformity it causes will result in fewer solutions to problems and less flexibility in confronting new solutions compared to a society that has many diverse institutions, despite the fact that the latter solution may appear to the population to contain inherently more disorganisation and lack of coordinated action.

As far as is possible the technical detail and management of society should be carried out by voluntary association [56]. The state should limit itself to setting the general laws and standards under which this technical work will be implemented; this may include legislation on standards of industries, licensing standards, and laws on the formation of associations and industrial organisation. Oxygen may be produced by a multitude of people and in many ways, but under a commonly agreed safety standard, just as water or gas is supplied in countries on the Earth by multiple corporations under strict state-guided water and gas quality standards. By focusing purely on generalities of legal process, the state can prevent itself from becoming drawn into limited special interests and specific political situations, which will inexorably cause it to slide towards intervention at many levels in a plethora of industries.

This general principle, as with many states on the Earth, also allows one to circumscribe the role of the police or similar security organisations within an extra-terrestrial society. Their job is to impartially uphold the general laws implemented by the judiciary and not to wield arbitrary power against the population or to act on the orders of isolated decisions and dictates from minorities or small bands of people holding power. The state will be regarded as a powerful and necessary monopoly on the use of violence. Given that criminality against infrastructure could threaten a large number of people in an extra-terrestrial environment, it is easy to see how the urgency that might be associated with acts to prevent potentially devastating criminality could be used to justify arbitrary police action (act now and prevent destruction, consider the consequences later). However, it will be a necessary culture within the extra-terrestrial state, just as it is in states threatened by terrorism on Earth, to avoid using these excuses to curb civil liberties and allow police (or their equivalent) to act with acts of random reprisals against criminal or potential criminal actions.

The second cause of a general will is likely to be the tendency of the state to attempt to achieve economic equality across society. Economic inequality is more likely to lead to criminality and an indigent subpopulation of people than on the Earth. In the confined environments of outer space both of these phenomena are a threat to the safety of other individuals and the economic and political order. On the Earth, poor people and criminals are more readily absorbed into large populations and are more easily accommodated within natural environments where basic commodities required for survival such as water and air are widely available. The extra-terrestrial state will develop a strong compulsion to quash these social evils by the forced redistribution of material resources to those who require them.

To a certain extent, such a compulsion should not be discouraged; criminality and poverty must be combated to the maximum extent possible, as they are on the Earth. But the state must resist the temptation to slide into a political system that seeks economic equality for all members of an extra-terrestrial society. One of the most efficacious bulwarks against the conformity that the environment will naturally impose upon society is a state and its associated culture that encourage individuals to express diverse economic and political ambitions that reflect their individual personalities. Those who seek riches or particular types of wealth (including scientific or artistic achievement[57]), however obnoxious some of these expressions may be to others, should not be discouraged from doing so [58]. Perhaps as importantly, these same people can become the powerful supporters and patrons of others, particularly those younger than themselves, who seek the same expressions of individual freedom. Thus a continuity and culture of independence of mind can be better secured by ensuring that in every generation there are those who have the intellectual and material means to challenge the extra-terrestrial state and other forms of governance and that they have the means to encourage others to do so. It is through economic diversity and inequality that extra-terrestrial society will find one rare method of achieving plurality in the human population.

The attempt to prevent equality of outcome must not be confused with a concomitant desire to accomplish equality of opportunity in certain fields. As elaborated in other sections of this essay, the prevention of cartels, monopolies and other special interest groups that will naturally tend towards the control and centralisation of extra-terrestrial society rests completely on the desire to open as many educational and political opportunities to all members of the population – itself encouraged by a strong state responsibility to promulgate a culture that seeks equality of opportunity [59]. The two fields in which equality of opportunity must surely be the highest priority are equality of opportunity to gain education and training (at whatever stage of life) and the equality of opportunity to seek information related to the running of extra-terrestrial political and economic organisations.

The most effective means to achieve economic inequality, but equality of opportunity in an extra-terrestrial society is to create a decentralised state which operates in as many of its decision making processes at a local level, allowing individuals to take part in political debate and discussion; and at the same time to construct an economic free enterprise system that allows individuals to achieve economic ambitions and to express economic visions that will lead to healthy economic inequality, but not poverty. To achieve the two highest priories in equality of opportunity there must be an extra-terrestrial education system that gives individuals the capacity to take part in society by maximising the opportunity for all the members of the society to learn the necessary skills and knowledge. Equality of opportunity in the extra-terrestrial environment must also be strongly predicated on the idea of equality of opportunity to collect information, political and economic, which reduces the possibilities for the state monopolisation of information that can be used in nefarious ways against the population. One obvious means by which this can achieved is the encouragement of free discussion through a maximally free press, a culture that induces people to openly promulgate ideas, contrarian and conservative, and the compulsory public availability of corporate and industrial data relating to finance and management.

The definition of the state in this way provides a foundation from which specific policies and institutions can be created. These specifics are historically contingent and determined by the needs and wants of the moment at which a solution is required. Once the general role of the state is understood, then there need be less concern that the vicissitudes of the moment will nurture tyranny since specifics will conform to the general liberty-seeking agenda of extra-terrestrial authorities as a whole.

To conclude, the purpose of this essay was not to provide an all-encompassing description and definition of the state, but instead to contribute to the long journey of drawing out, from an understanding of the way in which liberty might be expressed in extra-terrestrial environments, the general definition of institutions required to maximise extra-terrestrial liberty against its major threats and to minimise despotism for which the extra-terrestrial environment provides fertile ground.

Notes and References

1. Examples of such attempts include Plato's "*Republic*", More's "*Utopia*" and Bacon's "*The New Atlantis*". Few of these tracts are regarded as practically useful suggestions on how to direct or create real societies, although they have been the subject of debate on the extent to which their contents provide lessons on social organisation.
2. Of course, exactly this was done by Jefferson, Madison and the other framers of the American Constitution and its amendments; but they were designing a constitution for an existing and real population. There is a level of presumption that did not exist for them, but which does exist when trying to frame the structure of an extra-terrestrial state for people who do not yet exist and may have very different ideas about the route to freedom. To use the American analogy, to devise an extra-terrestrial constitution at the time of writing would be similar to a group of Europeans designing a state structure for the United States prior to constructing their first colony, and in this respect a certain degree of humility and caution must be exercised in recognising that the only true arbiters of the structure of the extra-terrestrial state must ultimately be the people who will live under it (a similar point was made in an unsurprisingly candid way by Paine in 1776: 'And when a man seriously reflects on the precariousness of human affairs, he will become convinced, that it is infinitely wiser and safer, to form a constitution of our own in a cool and deliberate

manner', T. Paine, "*Common Sense*", Penguin, London, 2004, p. 44). One might go further and declare it to be immoral and wrong for any individual and group not living in the environment in question to attempt to design a constitution (which is why the discussion here is limited to general observations about the limits of an extra-terrestrial state.)

3. C.S. Cockell, "An Essay on Extra-terrestrial Liberty", *JBIS*, **61**, pp. 255-275, 2008. This essay was primarily concerned with understanding how liberty would be expressed in extra-terrestrial environments and what threats it might face, although some sections did briefly allude to the types of institutions that represent its greatest perils.

4. These regions are not without food, however. The Inuit of the High Arctic hunt fish and seal. Prior to the International Antarctic Treaty it was not forbidden to eat the indigenous Antarctic wildlife, and this is an important distinction with extra-terrestrial environments where no indigenous food of any kind exists for people to gather. However, crops cannot be grown naturally in extreme polar environments, although experiments using greenhouses to grow food in the High Arctic have been undertaken and small crop growth units have been constructed in Antarctic stations.

5. Of course, there are many natural environments on the Earth with unbreathable atmospheres, including the interior of some caves. However, the distinction between these environments and extra-terrestrial environments is that people are not forced to live within these unbreathable extremes on the Earth, although they may choose to visit them for scientific exploration.

6. Throughout this essay I use the word 'state' to refer broadly to any form of extra-terrestrial governance and I include not merely the apparatus through which governance is exerted (such as the legal system), but also the government itself. For the sake of convenience within my definition I include the smallest committees that might oversee an extra-terrestrial outpost from a very early stage through to large fully formed states analogous to those that oversee entire nations on the Earth. Whereas in practice this may be an overly ambitious and simplified generalisation, it is useful here for discussing the general characteristics of the structures of governance in extra-terrestrial environments.

7. These states of nature have taken various forms. The most well-known expositions are Thomas Hobbes's (*"Leviathan"*, Oxford University Press, Oxford, UK, 1998 (first published in 1651)), John Locke's (*"Two Treatises of Government"*, Everyman Library,

New York, USA,1993 (first published in 1689)) and Jean-Jacques Rousseau's (*"The Social Contract"*, Penguin, Harmondsworth, UK, 1976 (first published in 1762)).

8. There seems no reason to believe that human nature will fundamentally change once we achieve a permanent presence in outer space, although certain perspectives on the Earth and society may undergo development and subtle revision, as has been the case during the development of society on the Earth. For example, there is no evidence that radical scientific developments, such as the theory of evolution or the solar-centric structure of the Solar System have changed human behaviour, a point briefly discussed in relation to space settlement in C.S.Cockell, "Visions of Ourselves", *Nature*, **457** p. 30, 2009.

9. The proliferation of these machines may itself dumb resistance to control: 'From the moment when the mind which has worked out a method of action has no need to take part in the job of the execution, this can be handed over to pieces of metal just as well as and better than to living members; and one is thus presented with the strange spectacle of machines in which the method has become so perfectly crystallized in metal that it seems as though it is they which do the thinking, and it is the men who serve them who are reduced to the condition of automata', S. Weil, *"Oppression and Liberty"*, Routledge, London, p. 88, 2002 (first published 1955). Much of the industrial infrastructure, particularly oxygen production, in extra-terrestrial societies will be in the immediate control of machines, with no precursor phase involving 'living members'. Weil's analysis strengthens the later conclusions about the importance of education as a means to prevent Weil's predicted outcome.

10. The vital requirement for oxygen lends itself most clearly to a system of monopoly that probably would come closest to a form of extra-terrestrial mercantilism. On the subject of terrestrial mercantilism Adam Smith had some of the most eloquent observations, e.g., 'A great empire has been established for the sole purpose of raising up a nation of consumers who should be obliged to buy from the shops of our different producers all the goods with which these could supply them. For the sake of that little enhancement of price which this monopoly might afford our producers, the home consumers have been burdened with the whole expense of maintaining and defending that empire … It cannot be difficult to determine who have been the contrivers of this whole mercantile system; not the consumers, we may believe, whose interest has been entirely neglected; but the producers, whose interest has been so carefully attended to; and among this

latter class our merchants and manufacturers have been by far the principal architects', A. Smith, "*An Enquiry into the Nature and Causes of the Wealth of Nations*", Everyman Library, New York, pp. 595–596, 1991 (first published 1776).

11. It might seem that an easy way to minimise the problem of oxygen monopolisation is to exclude it through a constitution. Superficially, the Fifth and Fourteenth amendments of the American Constitution which protect 'life, liberty and property' (within due process) would seem to prevent the usurpation of oxygen supplies by government organisations and their subsequent abuse as a form of coercion. However, these amendments are unlikely to be adequate for this task; private corporations can equally consolidate monopolistic power over oxygen supply outside the realm of government. But even if a Constitution was expanded to include private industry, the government and industry could invoke these amendments or their extra-terrestrial analogues as an excuse for control over oxygen, by convincing people that a reason for the control of oxygen was to protect their life and liberty from mechanical failings and control by the ever-present threat of tyrannies even worse than the existing arrangements. Through such arguments these amendments may be turned on their heads and become the very instruments of tyranny under the disguise of protecting individuals' constitutional rights. The only real solution to this entire problem is simply to maximise competition and minimise state and private monopolies in oxygen production, giving people the power to turn their backs on government and private organisations they perceive to be abusing oxygen production to exercise power.

12. The problem of the effects of economic centralisation on political freedom has been adequately dealt with elsewhere, but this comment by Simons particularly well sums up the problem with extra-terrestrial oxygen manufacture and other extra-terrestrial processes with involved technical requirements: 'Large governments, like giant business corporations, may effectively mobilise existing technology, realising fully its current potentialities. In a shortsighted view they are instruments of progress; but they lack the creative powers of a multiplicity of competitive smaller units', H. C. Simons, "*Economic Policy for a Free Society*", University of Chicago Press, p. 14, 1948.

13. Although, of course, war is equally an incentive towards centralisation of industry and political organisation. Indeed, Chomsky makes the case that military activity is a deliberate mechanism for state influence

even in times of peace, N. Chomsky, "*Government in the Future*", Seven Stories Press, New York, p. 59, 2005. The lack of war on other planetary bodies will remove one excuse for totalitarian states to centralise economic and political activity.

14. The problem of socialist economic calculation was elaborated by Ludwig von Mises (e.g. L. von Mises, "*Socialism: An Economic and Sociological Analysis*", Liberty Fund, Indianapolis, IN, 1981 (first published 1922) and "*Bureaucracy*", Liberty Fund, Indianapolis, IN, 2007 (first published 1944)). Hayek elaborates the three monumental data sets that would be required by the state to make these calculations. (F.A. Hayek, "*Individualism and Economic Order*", Chicago University Press, pp. 154–5, 1980 (first published 1948)). They comprise: knowledge of all products, the calculation of equilibrium within the system that would allow one to assign prices to these products, and knowledge of the relative importance of different goods to consumers. Although these data sets are quite impractical to obtain for nations involving millions of people, it might well be practical to gather them in small extra-terrestrial settlements of several hundreds to thousands of people. Although Hayek spends considerable energy examining the various problems that confront socialist central planning, he does recognise that the fact of whether it is technically achievable is rendered somewhat irrelevant by the greater concern of whether political freedom can ever be achieved in such a system: 'The question of how far a socialist system can avoid extensive central direction of economic activity is of great importance quite apart from its relation to economic efficiency; it is crucial for the question of how much personal and political freedom can be preserved in such a system' (p. 203).

15. Managers of Antarctic stations succeed in predicting demand from station inhabitants and planning for sufficient supply to cover those demands. In cases where demands outstrip supply they are also capable of moderating the supply of those consumables to meet demands in novel ways. In this way, successful small economies are run that are based on continuous calculations. Although in the case of small stations the system is one of direct and free supply according to basic needs, in the case of McMurdo Station, a US Antarctic station with over 1,000 inhabitants during the summer, money is used for some transactions, showing that a fledging economy can be constructed which is at the same time essentially centrally planned and coordinated.

16. The fear of lack of basic resources will encourage people to come together to coordinate production and centralisation of these

resources. But this fear must not overcome the equal desire to compete. A lack of resources will become a powerful political basis for consolidating centralisation in the interests of public safety, but worse still, for singling out those who engage in competition that might result, now and then, in unpredictable supply problems, of being responsible for threatening the lives of the public. The state and its officials will inevitably have to overcome the desire to use unpredictability that results from competition as an excuse to demand and implement the centralisation of resource production and allocation.

17. Von Humboldt, in his well known exposition of the limits of state power, states: 'inasmuch as the state, in its positive care for the external and physical well-being of the citizen ... cannot avoid creating hindrances to the development of individuality, we find another reason why it should not be permitted to exercise such interference except in the case of the most absolute necessity', W. von Humboldt, "*The Limits of State Action*", Liberty Fund, Indianapolis, IN, p. 28, 1993 (first published 1791). But the problem with this view, as discussed in a previous essay (Cockell, "An Essay on Extra-terrestrial Liberty"), is that the extra-terrestrial state itself can define the realm of actions that fit within the phrase 'absolute necessity' and thereby use this type of phrase as an instrument of positive liberty and thence tyranny. Where industries are vital to human survival, as is the case with oxygen production, then any state action pertaining to this industry, or even any industry vaguely related to the running of this core industry, can be justified as absolutely necessary.

Humboldt later makes the same type of observation when discussing state security: 'Disturbances of security are produced either by actions which in themselves violate the rights of others, or by those from whose consequences this is to be feared' ("*The Limits of State Action*", p. 85). It has been the preserve of tyrannies since the beginning of civilization to convince the population that they should fear this and that group or individual and thereby to generate the culture and rationale for state action or coercion. The problem in extra-terrestrial societies, as it is on the Earth, is to define the limits of 'absolute necessity' and when the state is justified in intervening in activities out of fear of negative consequences. It is clear that the risk to life and property in extra-terrestrial environments, caused by such actions as depressurisation of habitats, provides justification in favour of greater state action as a precautionary measure against the fear of impending disaster.

18. 'But from the point of view of the real, the historical roots of liberal democracy, freedom has rested neither upon release nor upon collectivisation but upon diversification and the decentralisation of power in society. In the division of authority and the multiplication of its sources lie the most enduring conditions of freedom. "The only safeguard against power," warned Montesquieu, "is rival power"', R.A. Nisbet, "*Community and Power*", Galaxy Books, New York, pp. 269-70, 1965.
19. In the terrestrial case succinctly stated as: 'In modern society, however, the essential requisite for the protection of the individual against coercion is not that he possesses property but that the material means which enable him to pursue any plan of action should not be all in the exclusive control of one other agent ... The important point is that property should be sufficiently dispersed so that the individual is not dependent on particular persons who alone can provide him with what he needs or who alone can employ him', F.A. Hayek, "*The Constitution of Liberty*", Routledge, London, pp. 123-4, 2006 (first published 1960).
20. The problem lies in the ambiguity of necessity versus gratuitous control. Malinowski states: 'When discipline is brought into being by a temporary inevitable crisis it must be accepted or else the group may perish. When discipline is imposed upon a community and the culture as a whole, transforming thus the whole group into a passive instrument of power politics, it destroys the very core of civilization', B. Malinowski, "*Freedom and Civilization*", Indiana University Press, p. 203, 1960 (first published 1944). However, when does control of the oxygen supply system shift from necessary control to avert a crisis to being a passive instrument of power? There is probably no clear-cut transition, but an obvious overlap in intentions. It is this overlap which provides the argument to remove oxygen supply from central state control and maximise competition, thus minimising its power over society and thwarting the temptation to usurp it as an instrument of power.
21. Fragmentation is not merely a way to prevent the consolidation of state power over industry, but also a means to discourage private industry itself becoming its own consolidated tyranny, recognising that private industry has just as strong a capability of becoming an agent of despotism as state influence: 'The love of power, the selfishness, the injustice, the untruthfulness, which often in comparatively short times bring private organisations to disaster, will inevitably, where their effects accumulate from generation to generation, work evils far greater and less remediable; since, vast and complex and possessed of all resources, the administrative

organisation once developed and consolidated, must become irresistible', H. Spencer, *"The Man Versus the State"*, University Press of the Pacific, Honolulu, HI, p. 331, 2003 (first published 1884).

22. The multiplication of offices has been one means by which tyrannies mask the true identity of responsible organisations, thus throwing the population into a form of helplessness about who to appeal to for a social compass. Thus, despite the fact that the multiplication of bureaucracy looks like an increase in social order and structure, it is in fact a means to reduce order and transform society into an amorphous mass. This fact was noted by Hannah Arendt in relation to Nazi society: 'As a matter of fact, duplication of offices, seemingly the result of the party-state problem in all one-party dictatorships, is only the most conspicuous sign of a more complicated phenomenon that is better defined as multiplication of offices than duplication … He [i.e. the public] had to develop a kind of sixth sense to know at a given moment whom to obey and whom to disregard', H. Arendt, *"The Origins of Totalitarianism"*, Harcourt, London, p. 399, 1994.

23. It is beyond the scope of this essay to discuss how the system of public engagement in the polity could be organised. The implicit assumption in this essay is that it is through democracy; if it is not, then extra-terrestrial society is likely to succumb quickly to tyranny. In the earliest stages of an extra-terrestrial society a truly active form of participatory democracy may be possible akin to Israeli kibbutzim. As the scale of a society increases it must surely transition to a model more similar to representative democracy. It might seem intuitive that representative democracy is more likely to fall prey to tyrannical influences as representatives may be corrupted by oxygen producers and other special interests, but there is no reason why even in a participatory democracy cabals could not dominate discussion.

24. 'It is not the powers which democratic assemblies can effectively wield but the powers which they can hand over to the administrators charged with the achievement of particular goals that constitute the danger to individual freedom today', F. Hayek, *"The Constitution of Liberty"*, p. 101, and later: 'It would scarcely be an exaggeration to say that the greatest danger to liberty today comes from the men who are most needed and most powerful in modern government, namely, the efficient administrators exclusively concerned with what they regard as the public good', p. 228. Hayek's comments concerned the government of the day (1960), when his book was published, but his observations succinctly describe the inevitable problem in extra-terrestrial

societies where the efficient and responsible administration of limited resources will be demanded by people.

25. A point that should be raised is that one could argue that if oxygen is produced and distributed in a modular fashion it might be easier to threaten individuals by switching off their supply, in a targeted manner. One could argue that the production of oxygen that is injected into common spaces is actually less likely to result in tyranny because the authorities will find it more difficult to threaten large numbers of people and get away with it. Whilst I see merit in this view, I believe, on balance, that a general culture of coercion and the likelihood of central tyranny is more likely to result from the common production of oxygen than an attempt to modularise production and create smaller entities producing it (as with water and food production).

26. A terrestrial response to this question can be found in Kingdon's summary: 'In the light of man's needs and of unfolding events [the Second World War], we are justified in declaring that education must mould its forms to serve four clear ends. It must train minds to think in cosmopolitan terms that will enable them to see local cultures against the corrective perspective of world experience and so fit them for the creative emergence of a comprehensive culture of mankind. It must maintain an active international fellowship of free minds meeting and sharing each other's research and discoveries in the assurance that facts and their implications know no bounds of parish or nation or race. It must courageously act upon its own premise that reason is man's most expert instrument for mastering the physical world and organising his own society, and when the processes of thought are threatened by the restrictive dogmatisms of any political forms it must at any cost refuse to surrender the independence of the mind. It must recognise that its full task consists not only in training people in freedom but also in preparing them for freedom, which means for control of themselves as well as of their world, and includes educating them consciously for change and, at the same time, for responsibility', F. Kingdon, "Freedom for Education", in *Freedom: Its Meaning*, ed. R.N. Anshen, George Allen and Unwin, London, p. 147, 1942. This statement sets a credo to follow, but it requires a further level of analysis to identify the specific problems that are to be found in the extra-terrestrial environment, particularly the imposition of the physical environment on young minds. Kingdon's concern about parochialism of local cultures may equally apply, but will be more problematic in space, since there will be no 'world culture' as such (at least surrounding an extra-

terrestrial settlement in the earliest stages of these societies), so that creating a cosmopolitan mind will require enormous internal efforts in society to generate variation of thought.

27. Hayek claimed that: 'the more one highly rates the power that education can have over men's minds, the more convinced one should be of the danger of placing this power in the hands of any single authority', and, referring to our increasing knowledge of how to direct the education of children towards specific ends, 'Indeed, we may soon find that the solution has to lie in government ceasing to be the chief dispenser of education and becoming the impartial protector of the individual against all uses of such newly found powers', *"The Constitution of Liberty"*, pp. 328, 329. Hayek's views, however (and those probably of many people today) were very different from those of educationalists of the eighteenth century who 'considered supervision of the establishment of habits and associations to be central to education', D. Spadafora, *"The Idea of Progress in Eighteenth Century Britain"*, Yale University Press, New Haven, CT, p. 174, 1990. It is easy to envisage how, confronted with the extremities of the extra-terrestrial environment and the potential consequences of criminality, educators and state planners in extra-terrestrial societies would revert back to a type of eighteenth-century conviction that the tight control and direction of human minds is the way to produce a safe and well-run society. Although this may well be the case in some elements of education (such as teaching children the basics of responsibility in enclosed environments and the need for selflessness), applied across all aspects of education it is a recipe for despotism.

28. The potential value of unpredictability and even danger was recognised by Tocqueville: 'I think, then, that the leaders of modern society would be wrong to seek to lull the community by a state of too uniform and too peaceful happiness, and that it is as well to expose it from time to time to matters of difficulty and danger in order to raise ambition and to give it a field of action', A. de Tocqueville, *"Democracy in America"*, Wordsworth Editions, Ware, UK, p. 317, 1998 (first published 1835).

29. Rawls sees no place for this type of 'perfectionism'; his view of justice is limited to a achieving a 'neutral' state, indeed his conception of justice (the 'original position') was devised to avoid people, and thus the state, taking up specific notions of good activities (J. Rawls, *"Political Liberalism"*, Columbia University Press, New York, 1993). By contrast, Joseph Raz does see a place for a more interventionist state (J. Raz, *"The Morality of Freedom"*,

Oxford University Press, Oxford, 1986). The Rawlsian idea of a neutral state is absolutely required in the extra-terrestrial environment where the state will tend towards tyranny and it should avoid intervention in special interests; however, it could see its objective to intervene more generally to enhance cultural and scientific education as a means to improve the ability of people to engage with the organisation of society.

30. The problem of achieving a society of free-thinking people through educational establishments was recognised by Herbert Spencer: 'All institutions have an instinct of self-preservation growing out of the selfishness of those connected with them. Their roots are in the past and the present; never in the future. Change threatens them, modifies them, eventually destroys them. Hence to change they are uniformly opposed. On the other hand, education, properly so called, is closely associated with change – is always fitting men for higher things, and unfitting them for things as they are. Therefore, between institutions whose existence depends upon man continuing what he is, and true education, which is one of the instruments for making him something other than he is, there must always be enmity,' H. Spencer, "*Social Statics*", University Press of the Pacific, Honolulu, HI, p. 66, 2003 (first published 1897).

Milton Friedman elaborated a voucher system for private education by which parents could choose schools in which to redeem vouchers for their children's education (M. Friedman, "The Role of Government in Education", in *Economics and the Public Interest*, ed. R.A. Solo, Rutgers University Press, New Brunswick, NJ, 1955). It is beyond the scope of this essay to describe the details of Friedman's system, but its basic ethos was to create a mechanism for introducing choice into education and to weaken the power of state bureaucrats over who is educated where. In the extra-terrestrial environment, education, from elementary schools though to adult vocational training, is the most enticing means for the state to achieve the control of society.

31. Consider the following statement by De Lolme: 'The English Constitution being founded on such principles as those we have just described, no true comparison can be made between it, and the Governments of any other states; and since it evidently insures, not only the liberty, but the general satisfaction in all respects, of those who are subject to it, in a much greater degree than any other Government did, this consideration alone affords sufficient ground to conclude, without looking farther, that it is also more likely to be preserved from ruin', J.L. De Lolme, "*The*

Constitution of England", Liberty Fund, Indianapolis, IN, p. 316, 2007 (first published 1771). Now although one can no doubt agree that the English constitution is fine, the unquestioning tone of this conclusion illustrates quite aptly the way in which, by living inside a system, one becomes dulled to objective critique. It is a particularly apposite example because it was written in the eighteenth century about a nation that would go far in extending liberty to new horizons over the next two centuries. If one has no foresight about where liberty might go, it is unsurprising that one would conclude that one's own present situation represents the apex of liberty.

32. Cockell, "An Essay on Extra-terrestrial Liberty", p. 267.
33. That freeing people intellectually can sometimes make them more nervous and less able to express their potentials than when they are given strict orders, which provides them with the security of a well-defined order, is to be accepted (and is, indeed, the very credo of a totalitarian state). Whereas this may be the case for many people, the common weal is better served by seeking maximum freedom to allow those who gain from this type of environment to thrive and to minimise the emergence of a collective tyranny organised by those who feel safer under regulations. Chesterton's characteristically contrarian observation, although probably true, is a vision of totalitarianism if it is accepted as the basis from which to plan a society: 'We may say broadly that free thought is the best of all the safeguards against freedom. Managed in a modern style the emancipation of the slave's mind is the best way of preventing the emancipation of the slave. Teach him to worry about whether he wants to be free, and he will not free himself … the little clerk in Mr. Gradgrind's office – he is too mentally worried to believe in freedom. He is kept quiet with revolutionary literature. He is calmed and kept in his place by a constant succession of wild philosophies', G.K. Chesterton, *"Orthodoxy"*, Ignatius Press, San Francisco, p. 114, 1995 (first published 1908).
34. The 'tyranny of the majority' may equally come from the democratic interpretation of this phrase – tyranny resulting from the majority view imposed by election – or from conformist majority views that emerge from a consumer society where people's aspirations are levelled by ideas of fashion or the culture that prevails, particularly from advertising. Fromm observes: 'We forget that, although freedom of speech constitutes an important victory in the battle against old restraints, modern man is in a position where much of what "he" thinks and says are the things that everybody else thinks and says; that he has not acquired the

ability to think originally – that is, for himself – which alone gives meaning to his claim that nobody can interfere with the expression of his thoughts', E. Fromm, "*The Fear of Freedom*", Routledge, London, p. 91, 2006 (originally published 1941).

35. This point is revisited many times by John Trenchard and Thomas Gordon in their letters ("*Cato's Letters*") written between 1720 and 1723, but perhaps most bluntly stated in letter no. 14 (entitled *'The unhappy state of despotick princes, compared with the happy lot of such as rule by settled laws. How the latter, by abusing their trust, may forfeit their crown'*) published on 28 January 1721 in *The London Journal*: 'A good man will choose to live in an innocent obscurity, and enjoy the internal satisfaction resulting from a just sense of his own merit, and virtue, rather than aim at greatness, by a long series of unworthy arts, and ignoble actions; whilst the ambitious, the cruel, the rapacious, the false, the proud, the treacherous part of mankind, will be ever thrusting themselves forward, and endeavouring to sparkle in courts, as well as in the eyes of the unthinking crowd; and, to make themselves necessary, will be continually either flattering or distressing princes.' Letter no. 43, published on 2 September 1721, *'The Natural Passion of Men for Superiority'*, elaborates this view in more detail.
Perhaps a slightly more extreme view is enunciated by Bury: 'But the development of human societies has not been guided by human reason. Men have not consciously made general happiness the end of their actions. They have been conducted by passion and ambition and have never known to what goal they were moving. For if reason had presided, progress would soon have been arrested', J.B. Bury, "*The Idea of Progress: An Inquiry into its Origin and Growth*", Echo Library, Teddington, UK, p. 83, 2006 (first published 1920).

36. Although I think this view is essentially robust, it has not been unknown for powerful religious groups to mitigate state tyranny. As De Ruggiero points out: 'Ever since the feudal period, and with renewed vigour in the age of absolute monarchy, the Catholic Church has fought against state supremacy; and the very fact of this conflict between two great powers has been an effectual safeguard for individuals against the peril of utter enslavement to either', G. De Ruggiero, "*The History of European Liberalism*", Beacon Press, Boston, p. 19, 1966. However, to rely on the hope that a powerful religious group would end up in conflict with the extra-terrestrial state, thus mitigating tyranny, rather than merging with the state, would obviously be a foolish gamble. De Ruggiero later observes, in a discussion about religious mobility in England:

'Religion thus acquires the function of a powerful stimulant to the circulation of social energy, and a bond uniting class to class. In its hierarchical distribution, no less than in the special characteristics of its individual form, it makes at once for stability and for progress', p. 122–3.

37. '… every state-church is essentially popish', H. Spencer, "*Social Statics*", p.143.

38. Any attempt to remove such a tyrannical influence may itself be seen as a threat to what freedom is remaining. This represents the perpetual problem of societies in which tyranny has become the guardian of certain other defended freedoms. In examining the failure of democracy to adequately replace monarchy in Germany, Bernstein observes: 'This lesson must be drawn: that progress in human freedom cannot be made, that restraints to political or economic or cultural freedom cannot be removed, if these restraints satisfy cultural and political needs which are imperative, and contribute to the stability of society', F. Bernstein, "The Balance of Progress of Freedom in History" in "*Freedom: Its Meaning*", ed. R.N. Anshen, p. 47. However, this seems a somewhat specious lesson because on such a basis any tyranny could defend some token freedoms and then claim that the removal of its political system would undermine human freedom since the system was vital to stability. The point must be that although the removal of tyranny may threaten short-term freedom, in most cases it must be the case that this is worthwhile to secure a greater long-term freedom. Nevertheless, if religion becomes entrenched in an extra-terrestrial society and evolves into a form of despotism, it will become easy for religious leaders to provide compelling evidence that against the lethal extra-terrestrial environment religious devotion provides safety, cohesion, and therefore vital stability to society, the enemies of which are a problem to society.

39. M. Weber, "*The Protestant Ethic and the Spirit of Capitalism*", Routledge, London, 2007 (first published 1904).

40. ' The idea that religious liberty is the generating principle of civil, and that civil liberty is the necessary condition of religious, was a discovery reserved for the seventeenth century. … The cause of religion, even under the regenerate influence of worldly passion, has as much to do as any clear notions of policy, in making this country the foremost of the free', J.E.E. Dalberg Acton (Lord Acton), "*The History of Freedom and Other Essays*", Cosimo Classics, New York, pp. 52, 53, 2005 (first published 1922).

41. Von Humboldt states: 'In the endeavour to act upon morality through the medium of religious ideas, it is especially necessary

to distinguish between the propagation of a particular form of religion, and the diffusion of a spirit of religiousness in general. The former is undoubtedly more oppressive in its character, and more harmful in its consequences; but, without it, the latter is hardly possible. For when once the state believes morality and religiousness to be inseparably associated, and considers that it can and may avail itself of this method of influence, it is scarcely possible not to take one religion in preference to any others under its protection, according to its conformity to true, or generally accepted, ideas of morality', W. von Humboldt, "*The Limits of State Action*", p. 54.

42. Malinowksi observes: 'Insofar as property means an exclusive access to the use of certain instruments and the ability to deprive others of the use of such instruments or of the benefits derived from such a use, we have potentialities of slavery in the economic sense', B. Malinowski, "*Freedom and Civilization*", p. 251. It is interesting to note that Malinowski's antidote for this problem is that: 'education, justice, religion, and economic enterprise must remain largely independent while they also remain open to inspection', p. 251.

43. From a speech in the House of Commons, London, November 1783.

44. Hayek observes that: 'What our generation has forgotten is that the system of private property is the most important guarantee of freedom, not only for those who own property, but scarcely less for those who do not. It is only because the control of the means of production is divided among many people acting independently that nobody has complete power over us, that we as individuals can decide what to do with ourselves ... Who can seriously doubt that a member of a small racial or religious minority will be freer with no property so long as fellow members of his community have property and are therefore able to employ him, than he would if private property were abolished and he became owner of a nominal share in the communal property', F. Hayek, "*The Road to Serfdom*", Routledge, London, p. 108, 2007 (first published 1944).

45. These discussions include: P.F. Uhlir and W.P. Bishop, "Wilderness and Space", in *Beyond Spaceship Earth*, ed. E.C. Hargrove, Sierra Club Books, San Francisco, pp. 183-210, 1986; I. Almár and A. Horváth, "Do We Need 'Environmental Protection' in the Solar System?", in *Space Safety and Rescue*, ed. G.W. Heath, American Astronautical Society, Springfield, VA, pp. 393-8, 1989; I. Almár, "What Could COSPAR Do to Protect the Planetary and Space Environment?", *Advances in Space Research*, **30**, pp. 1577-81, 2002; C.S. Cockell and G. Horneck, "A Planetary Park System

for Mars", *Space Policy*, **20**, pp. 291-5, 2004; C.S. Cockell and G. Horneck, "Planetary Parks – formulating a Wilderness Policy for Other Planetary Bodies", *Space Policy*, **22**, pp. 256-61, 2006.

46. C.S. Cockell, "A Simple Land Use Policy for Mars", in J.D.A. Clarke, *Mars Analog Research*, **Vol. III**, American Astronautical Association, Springfield, VA, pp. 301-11, 2006.

47. *"United Nations Treaties and Principles on Outer Space"*, United Nations, New York, 2002. The UN Outer Space and Moon Treaties declare that extra-terrestrial real estate "shall be the province of all mankind". White and Dunstan (In *"Space-The Free Market Frontier"*, Cato Institute, Washington DC, USA, pp. 83-124 and pp. 223-241, 2002) discuss the implications of the treaty and an interpretation that allows for private ownership of resources, but precludes ownership of entire planetary bodies, particularly by nations. Whatever the interpretation of the treaty and the possibility for follow-on treaties to enshrine private ownership, the denial of the possibility of owning private land on other planetary bodies under any derivation of the basic idea of extra-terrestrial bodies being the province or heritage of humankind is tyrannical in that it subjugates individuals to the collective will of humankind, which is likely to be so diffuse and lacking in common agreement that it consigns extra-terrestrial land to an unproductive limbo. In that land offers people resources to achieve survival and may be necessary to develop a basic subsistence in space, then it offers the most likely means for achieving a permanent human presence in new areas.

48. The suggestions in this essay are inspired by traditional solutions to the problem of mitigating tyranny, particularly the encouragement of individual autonomy, for example the general approach by Mill (J.S. Mill, *"On Liberty"*, Oxford University Press, Oxford, 1998 (first published 1869)). Nevertheless, there is a distinction between the extra-terrestrial and terrestrial case. In both cases individual autonomy is suggested as a means to mitigate state tyranny by giving the public the education and intellectual tools to think for themselves. However, in the terrestrial case it tends to be regarded as a beneficial contribution to promoting liberty. In the extra-terrestrial case, because the environment creates a cryptic natural tyranny, there is a degree of essentialness to this task that is lacking in the terrestrial case. Without encouraging independence of mind and free thinking, the extra-terrestrial society will decline into a dictatorship with a greater degree of certainty than is the case on the Earth.

49. The requirement is not removed in a democratic system. Quite apart from the 'tyranny of the majority', an old problem with

democracy, Tocqueville recognised that democracy inherently tends to create a population more concerned with everyday affairs: 'I think that it is extremely difficult to excite the enthusiasm of a democratic people for any theory which has not a palpable, direct, and immediate connection with the daily occupations of life; therefore they will not easily forsake their old opinions, for it is enthusiasm that flings the minds of men out of the beaten track and effects the great revolution of the intellect as well as the great revolutions of the political world', A. de Tocqueville, "*Democracy in America*", p. 329. Such a view leads to the paradoxical conclusion that democracy may even strengthen the ease with which the extra-terrestrial environment causes conformity to set into society and cause even greater efforts to be required from people and institutions to excite enthusiasm for unconventional ideas.

50. Rousseau (J-J. Rousseau, "*The Social Contract*", Penguin, Harmondsworth, UK, 1976 (first published in 1762)) views the 'general will' as that part of the human psychology which engages with thoughts and activities that are for the benefit of the common weal. A free individual is one whose will coincides with this general will. If it does not, then the state can legitimately force an individual to change their behaviour to coincide with the general will; in this way they are 'forced to be free'. If this line of logic might appear to be an abomination to a modern mind, it should be remembered that in the 18th Century it was regarded as a startling new insight into the nature and conditions required to secure liberty. Since the extra-terrestrial environment is lethal then this same line of logic can be easily justified in space. A person who disagrees with the general will of society and the state could be portrayed amongst their peers as a person who is not only a threat to others but has become deranged and a danger to themselves. To secure the protection of both the public and the "miscreant" from death the state could legitimately intervene to quash their dissent and thereby 'force them to be free' (free from their errant ideas which will "inevitably" lead to their death and a danger to others). Rousseau's vision of the state is open to appalling abuse in the extra-terrestrial environment.

51. Thus, even in a democracy, people will tend to unite into a common front against the extra-terrestrial environment, and this will breed uniformity. The basic principle of how parties form is stated by Ferguson: 'In every casual and mixed state of the national manners, the safety of every individual, and his political consequence, depends much on himself, but more on the party to which he is joined. For this reason, all who feel a common interest, are apt to unite in parties; and, as far as that interest

requires, mutually support each other', A. Ferguson, *"An Essay on the History of Civil Society"*, Echo Library, Teddington, UK, p. 124, 2007 (first published 1767). If safety is taken in its literal physical sense, and safety against the environment is seen to be a focus for concern, then people will tend to gravitate to parties that guarantee that protection, and political groups or individuals that preach the usurpation of vital industries that are necessary to provide that guarantee will make good headway.

52. On the subject of the general will, De Ruggiero observes: 'Thus the democratic state is the result of depriving the citizens of their rights and conferring them upon a general will, a single and indivisible sovereign people ... If it is to be a genuine fact, the general will and the will of all must exactly coincide; that is, the collective interest must be the arithmetic sum of the individual interests,' G. De Ruggiero, *"The History of European Liberalism"*, p. 375. Although this may appear implausible to some people, if there is a common focus for a society which seems undeniably common to all – for example, survival against a lethal extra-terrestrial environment – then it does not seem at all improbable that a general will would emerge that is completely in coincidence with the will of individuals, that is, the will to live safe lives without the threat of death from the instantaneously lethal environment. It is this very real possibility of the indistinguishable overlap of the general and individual will that the extra-terrestrial state must be charged to guard against, both as an end in itself and as a means to vitiate the abuse of an emergent general will as a tool of tyranny by those governing society.

53. The state depends upon the general support of the people for its existence, but garnering this support requires some type of ideology with which to persuade people to buy into the state. Achieving the balance of an ideology that binds the people's support, but prevents collapse, whilst at the same time preventing the ideology from becoming tyranny is the problem that must be addressed by both the state and the people together. 'This support need not be active enthusiasm to be effective; it can just as well be passive resignation. But support there must be. For if the bulk of the people were really convinced of the illegitimacy of the state ... then the state would soon collapse to take on no more status or breadth of existence than another Mafia gang. Hence the necessity of the state's employment of ideologists', M.N. Rothbard, *"The Ethics of Liberty"*, New York University Press, p. 169, 2002.

54. It is hardly necessary to point out that in addition to these measures, the involvement of as many people as possible in state

affairs should be encouraged. Godwin concisely lists a series of arguments for the involvement of many people in government which apply equally well to the extra-terrestrial state: 'Firstly, it has already appeared that there is no satisfactory criterion marking out any man, or set of men, to preside over the rest. Secondly, all men are partakers of the common faculty, reason; and may be supposed to have some communication with the common instructor, truth. It would be wrong in an affair of such momentous concern that any chance for additional wisdom should be rejected; nor can we tell, in many cases, till after the experiment, how eminent any individual may be found in the business of guiding and deliberating his fellows. Thirdly, government is a contrivance instituted for the security of individuals; and it seems both reasonable that each man should have a share in providing for his own security; and probable that partiality and cabal will by this means be effectually excluded. Lastly, to give each man a voice in the public concerns comes nearest to that fundamental purpose of which we should never lose sight, the uncontrolled exercise of private judgement. Each man will thus be inspired with a consciousness of his own importance, and the slavish feelings that shrink up the soul in the presence of an imagined superior will be unknown', W. Godwin, "*Enquiry Concerning Political Justice*", Penguin, Harmondsworth, UK, pp. 231-2, 1985 (first published 1793).

55. This would not be a new phenomenon. As Nisbet observed concerning agricultural practices in medieval society: 'To be sure there were variations in the intensity of this communality from one area to another, but wherever open-field husbandry was practiced the sheer technical demands of the system, with its complicated network of strips, enjoined upon the peasant a degree of solidarity with his fellows that the later enclosure acts and reform programs found difficult to break. The villager had little alternative, in such surroundings, but to subordinate himself and his desires to those of the village group', R.A. Nisbet, "*Community and Power*", p. 82. Within his analysis is thus the recognition not only that technical complexity and human interdependence tends to solidify conformity, but that once a system encourages this culture, it is difficult to remove or change.

56. A forthright view of the role of government and its relationship to voluntary association is expressed by Narveson: 'And the only reason why modern governments in the more decent places in the world are tolerable at all is because they approximate to some degree the model of free men and women working in concert or individually to get things done that they want done'. Narveson's

view is that without this character governments are 'to greater or lesser degrees incompetent', J. Narveson, *"The Libertarian Idea"*, Broadview Press, Peterborough, ON, p. 235, 2001.

57. Kelley asserts: 'The primary measure of a good society is the scope it affords achievement – the freedom society allows the able, the ambitious, and the productive to create value – value of any kind, from the production of material wealth, to the discovery of scientific knowledge, to the creation of art', D. Kelley, *"A Life of One's Own: Individual Rights and the Welfare State"*, Cato Institute, Washington, DC, p. 95, 1998.

58. If this seems a contradiction in that inequalities are likely to encourage the emergence of powerful individuals more likely to abuse this power (for example, the expression of economic influence through the control of the oxygen industry), I assume that the safeguards described earlier in the essay (corporate transparency, fragmentation of organisations where necessary) will minimise this danger. The tenet expressed here is that an extra-terrestrial society of total equality is more likely to feed conformity and ultimately cause tyranny than a society where there is inequality in individual economic capacities, even despite the concern that the latter arrangement is more likely to lead to the abuse of economic power by some individuals that are better off than others.

59. However, even complete equality of opportunity is clearly unachievable in practice. For example, concerning the way in which families, particularly parents, can create opportunities for children: 'Economic opportunity, in short, will never truly be equal so long as the private family exists in some form, as it should. While we should strive to make opportunities more equal, we must at the same time recognise sensible limits to this ideal', C. Duncan, "Democratic Liberalism: The Politics of Dignity", in *"Libertarianism: For and Against"*, eds C. Duncan and T.R. Machan, Rowman and Littlefield, Lanham, MD, p. 105, 2005. It seems that economic inequality and some inequality of opportunity must be accepted by the extra-terrestrial state and by the people, and the design of extra-terrestrial society adjusted according to this reality.

Liberty Across Light Years – The Freedom of Future Space Settlers Compared to that of the Ancients and the Moderns

There is a distinction between the liberties expressed by the ancients and that of modern states, a separation made clear by Benjamin Constant[1] when he admonished the protagonists of the French Revolution for their pursuit of political liberty at the expense of individual liberty.

The citizens of Sparta and Rome had little concept of the liberty of the individual. Decisions on the fate of the state and its people were considered subjects for collective deliberations and the prerogative to implement these decisions made individual freedoms subordinate. Perhaps the unusual, but nevertheless remarkable, exception to the rule was ancient Athens, but even there, the expression of individual views during the regular meetings of the Assembly – the gathered masses of eligible Athenians – left little room for private lives, for the pursuit of individual notions of the good life.

Modern states are so large, the apparatus of government so diffuse, that the direct participatory democracy of Athens has necessarily given way to representative democracies. In such populous states, where the power of the individual to influence society is so diminished, individual freedoms have become paramount in ensuring that the individual feels any self-worth at all. And so on a planet where the resources needed for survival; air, water and food, are generally plentiful, at least in western states at the time of writing, individuals can pursue their own lives and demand a minimum of interference from the state, such that the protection of individual freedoms and rights define the very essence of the operation of the state.

This difference in conceptions of liberty between ancient peoples and the moderns may not be as stark and well-defined as this description would suggest, and as Constant would have led us to believe. There was doubtless a sense of personal freedoms in some ancient societies, and today, of course, there are still examples of societies where individual freedoms remain secondary to the will of the state. However, in the broad sweep of history, the type of liberty reflected in ancient social arrangements, where the survival and persistence of relatively small groups of peoples was the priority that overshadowed any notion of the sovereignty of the individual, was a very different type than that which emerged in the modern nation-state, where the rights of human beings as distinct and separate entities are considered the precarious objects facing danger and the core of the very notion of liberty.

In space, lethal environmental conditions necessitate safety checks and controls and levels of collective responsibility that far exceed those required in most environments on the Earth. The relative isolation of economic systems in space, which lessen the opportunities for vigorous commerce, so beneficial to private property, and the small population sizes that make individuals less anonymous, create all the conditions for a type of liberty that comes closer to the liberty of the ancients than the moderns. It is difficult to conceive of any place in space where the political liberty of the collective, and its power to subordinate the individual, will not be seen as vital for the survival of a settlement. Individual expressions of freedom will easily become brash, indulgent and selfish expressions of people who appear to care little for the responsible alignment to group objectives, the implementation of which, with predictable and steadfast resolve, will be necessary for the continuity of an outpost.

Space settlers may be forced to adopt a type of liberty more akin to Sparta and Rome. This type of freedom might superficially appear to be a backward step, but if there is no choice in the matter and the environment of space requires this adjustment, then we may need to recognise that the freedom of modern terrestrial states is but one type of liberty appropriate to Earth. Outer space may require from people a complete reconsideration of the notions of liberty, the place of the individual and the importance of political liberty – a newfound awe for the collective will of society.

However, rather than resigning future space settlers to the liberty of the ancients, it might instead be timely to consider a new type of liberty in which the freedom of a group of people, or society, to ensure its own survival has primacy, but where mechanisms exist for protecting certain freedoms of the individual, where their powers to lead private lives are ring-fenced from the necessarily powerful forces of the collective political authority of a space settlement. We could arrive at a conclusion that space settlers will likely return to the liberty of the ancients, but, by learning the lessons of modern states, they might fashion means to ensure that individual liberties, the liberties of the moderns, are also protected.

It seems likely, therefore, that outer space offers the potential for a significant and quite remarkable historical transition in the history of human liberty. It might be the stage of synthesis in a type of dialectic. The thesis of liberty pursued by the ancients (collective liberty as the core of freedom) has given way to the antithesis of the moderns in which individual liberty takes priority over the liberty of the community. In space, a synthesis will occur as the political liberty of the group underpins the success of space settlement, like the ancients, but the liberty of the individual is protected as far as possible within it, as an overt objective of society, like the moderns. Lessons from the past and present will merge. So we should not view the future of liberty in space as merely another development or derivation of freedom, but rather a new and distinctive phase in the maturation of the concept of liberty, a synthesis of all our previous experience in which a balance between the group and the individual might be reached in a manner and with a character that has not previously been seen in any state, ancient or modern.

As a greater number of states and private companies propel themselves into space, the nature of extra-terrestrial liberty, and the mechanisms by which tyranny is to be thwarted, have become questions of imminent significance. Although a clear case can be made that freedoms must be constructed by extra-terrestrial settlers themselves as their social experiment emerges, allowing them to optimise the conditions for freedom as they see fit and most appropriate to the challenges they face, it must surely be a worthy thing for even terrestrially-bound people to consider the fate of freedom in the far flung future of space settlement? By considering the future of liberty beyond the Earth, a person need not be impinging on the self-determination of space settlers, but rather contributing to the sum of thought from which these settlers may draw their own conclusions. A discussion of liberty in outer space is an important objective for political philosophers and merits an invigorated discourse on the very structure of freedom, on Earth and beyond.

Charles S Cockell,
Edinburgh 2011

1. Constant, B., "The Liberty of the Ancients Compared to that of the Moderns", in B.Constant (ed.) Political Writings. Cambridge: Cambridge University Press, pp. 307-328, 1988.
2. Hopkins, M.M., "The Economics of Strikes and Revolts During Early Space Colonization: a Preliminary Analysis", Rand Corporation Document P-6324, 1979; Young, F.C., "Labor Relations in Space: An Essay in Extra-terrestrial Business Ethics", The Monist **71**, pp. 114-129, 1988.

On Tyranny

In the first essay I explored what I thought were the characteristics of liberty. In the second essay I investigated a little more of what I thought the implications of that might be for the structure of the extra-terrestrial state. But I was interested to think a little more about the sources of extra-terrestrial tyranny and to try to identify the sources of tyranny in space. It seemed to me that tyranny was not only a possible danger to individuals and societies in space, but also, ultimately, a potential threat to freedom on the Earth. It was one thing to discuss tyranny and its implications for social arrangements, but if extra-terrestrial societies are to try to avoid it, it is useful to identify where it might come from.

The Causes and Consequences of Extra-Terrestrial Tyranny

The construction of societies in space in which liberty can be preserved requires that the reasons for the emergence of despotism are identified. Tyranny will emerge from the historical origins of extra-terrestrial society and the way in which early communities must be developed technically. It will receive encouragement from the imposing nature of the extra-terrestrial environment – its extreme physical characteristics and vast spatial scales that encourage social isolation and autarky. It will flourish in the very culture of a new society in which the laws of physics force people to engage in the most traditional forms of revolutionary activity, such as inventing new calendars. Preventing the emergence of tyranny will not merely be essential for the freedom of people in such societies: the continuity of liberty on the Earth may depend ultimately upon the successful propagation of liberty in space.

1. Introduction

To predict the social structures, customs and institutions most likely to threaten liberty in the future is not an easy task, and any attempt to do so will almost certainly be to some degree erroneous. However, it is not the case that such an analysis is futile, because any enquiry of this type is likely to provide a starting point from which to understand the possible future sources of despotism. It may even assist in providing a method of identifying, in advance, the nature and origin of these problems and how to counter them.

The task is particularly difficult in the case of outer space. The nature of societies in space will be very different from those that we have known on the Earth. Fundamental human behaviour is unlikely to be radically changed by the environment of space, but the personal attributes of people that are brought to the fore and exert an influence on the type of society that develops are likely to be unique in their combination, just as every society on the Earth exhibits a certain character. On the Earth, these manifold and protean characteristics are forged by the history of each particular society or the civilisations within which they have been embedded, together with the physical environment in which they sit and the contemporary institutions that surround them [1].

It will not do, therefore, to try to understand the conditions for liberty in space by merely assuming that the past four hundred years of political philosophy and human history provide us with everything we need to know [2]. Instead, it is necessary to embark on a new investigation of how liberty is likely to be expressed in extra-terrestrial environments and what types of institutions should be created to maximise its chances for survival. Of course, any such enquiry can be assisted by drawing upon the hard won lessons obtained on the Earth; and there is much literature from which these previous ideas can be drawn.

During the seventeenth century, as the nation-state emerged, the focus of early political philosophers such as Hobbes, Locke and Rousseau was on the nature of political authority and obligation [3]. As the nation-state became an accepted reality of social order, so these discussions fell away, leading more contemporary philosophers to focus on the conditions for justice within the structure of the nation-state [4]. An enquiry into extra-terrestrial liberty must begin again with the question of political authority and the legitimacy of the extra-terrestrial state [5].

The case of extra-terrestrial liberty provides a particularly troubling challenge. On the Earth, any attempt to predict the future of freedom in a given society begins at least with knowledge of the accepted facts of that society, its history, culture and institutions. Although predicting the trajectory of a society within the milieu of this complex collection of information is a challenge reserved for those with a sanguine and spirited mind, the problem becomes immense, and some would say insuperable, for an extra-terrestrial society that does not yet exist, the institutions of which are unformed and which, as yet, has no history.

Despite the rather negative prognosis that might face anyone with the presumption to attempt this task, there are characteristics about the environment of outer space that make it a place very distinct from the Earth, the most obvious being the lack of breathable atmospheres, indigenous food and easily acquired liquid water. The absence or paucity of these three most vital resources required for human survival will force society to take on characteristics that no institutional resolve can overcome [6]. From this observation one can make the claim that the general trajectory of society in outer space is predictable, although as the vicissitudes of history and institutions at any given time dictate and impose themselves on society, so it will take a flexuous path along this overall arc [7]. Now, if this assumption is correct, and I will assume that it is, then it is possible to make predictions about the future of liberty in outer space and the source of its most serious threats, provided one avoids trying to predict details, which are likely to become more inaccurate as attempts to circumscribe their specifics are intensified.

Bearing in mind these general conditions, this essay is an attempt to identify the major general challanges to liberty in space and their sources. The intention is to expand upon the conclusions drawn in a previous essay [6] in order to form a more complete description of the sources of extra-terrestrial tyranny, and to attempt to derive a synthetic insight into the general character of extra-terrestrial despotism. Before investigating possible sources of tyranny in outer space, it is first necessary to identify the reasons that this enterprise is of any importance at all.

2. The Consequences of Extra-Terrestrial Tyranny

The ways in which a generally tyrannous state of mind might consolidate itself within extra-terrestrial authorities are as diverse as they are difficult to predict, but an important question to answer is whether, if these types of despotic authorities do emerge, they have any significance beyond the people who are directly affected by them.

The question is important to consider because if it is the case that tyranny will influence citizens beyond the confines of the settlements in which it has taken root, then there are reasons to try to prevent it that run deeper than merely sparing isolated groups of people the barbarism that might result from the behaviour of states that they have allowed to become dictatorial.

Despotism can be infectious and a concern might be that in allowing tyranny to take hold in one outpost it will thereby influence others. An interest in understanding tyranny might be motivated by the desire to prevent the transmission of a social disease through space [8]. This tendency for dictatorship to spread is often the result of the inclination of politicians to copy existing arrangements out of a sense of expediency. Usually, a lack of imagination or simply the desire to opt for a less risky, yet sometimes tyrannical, approach leaves politicians to revert to the *status quo*. Having said that, it must be recognised that in some cases extremely tyrannical and dynastic authorities may indirectly encourage other settlements or founders of other settlements to take a radically libertarian tangent in order to escape the perceived repression experienced elsewhere.

Outposts may take tyrannical paths if they are run by the coercively inclined who, being surrounded by other tyrannous settlements, use this as an excuse to implement the same types of governance. These authorities can provide a few token freedoms to create the illusion of a society more free than alternatives.

And the infectious nature of tyranny can overwhelm the good-intentioned. Even if an outpost is run by those inclined to liberty, it may be difficult for them to achieve a high degree of freedom if the communities with which they trade, or even those from which they receive vital resources in the same region, are run by despots. This situation amounts to little more than the obvious observation that a large collection of autarchic settlements will tend to drag new settlements into a similar frame of mind.

I think, though, that there is a much more profound implication which is a concern in the much longer term, and that is the influence of extra-terrestrial tyranny on terrestrial liberty. If space does become home to a collection of settlements, then there are several ways in which their decline into serfdom might ultimately threaten the conditions for terrestrial liberty. It has long been recognised that outer space is a strategically important position [9]. Anyone situated in the orbits or free spaces beyond the Earth looks down on the Earth-bound population. Like defenders on the top of a hill who stare down at their adversary in the valley below, there is a virtually unchallengeable psychological advantage to being located on the high ground. Strategically, outer space is advantageous because with low energetic requirements to move around the Solar System and acquire its vast resources, energetic and mineral, those who have a commanding, or even just an influential position, in this frontier essentially wield power in the infinite spaces beyond the Earth, which is, within this picture, a tiny isolated rocky surface at the bottom of one particular gravity well.

From this perspective it is not a difficult task to understand the influence of extra-terrestrial tyranny on terrestrial liberty. The Earth would labour under the threat and thought of tyrants who stare down from the top of the gravity well. The presence of despotic regimes in outer space will eclipse the Earth with the shadow of autocracy. They have every chance of plunging the

Earth into a new type of benighted age. This influence neither requires a large number of these settlements, nor that they even exert a considerable economic leverage on the Earth; it merely requires their presence in space and their opposition to the philosophy of liberty. Their existence will make those on the Earth who defend the arguments of liberty-seeking states feel insecure and vulnerable.

There are more tangible ways in which tyrants might influence the Earth. If they, at the top of the gravity well, develop weapons or even threaten to develop weapons that can be dropped onto the Earth, substantial economic resources might be expended by terrestrial states attempting to contain or prevent malevolent intentions from them. Like a person harassed by flies in the summer heat, they do not need to be large or even to possess the capacity to do much harm; they may be orders of magnitude smaller in resources and numbers of people than the Earth, but their presence might still be a constant source of irritation to earthly states. While they exist, the time and resources applied to them will be a drain on terrestrial liberty-seeking states. Their access to the infinite resources of space and their ability to secrete away weapons in the unpoliceable vastness of the interplanetary void will make them a persistent concern to terrestrial nations because their unknown intentions and future potential lead them to sap resolve and draw political and economic attention.

In the much longer term, these tyrants may control the supply of essential materials to Earth. Rare metals, sources of energy and key orbits in space may all eventually fall under their control. Commanding the vast resources and key strategic points of the Solar System, autocractic extra-terrestrial states do not need to possess expansive infrastructure to find themselves in a situation where they control important elements in the means of terrestrial supply.

All of these factors conspire to allow extra-terrestrial communities to exert a disproportionate effect on the Earth compared to their size and populace. It is not an over-dramatisation to conclude that if liberty should be extinguished in space then its future on the Earth will be in doubt. Pope Gregory VII's invocation to 'Let the terrestrial kingdom serve – or be the slave – of the celestial' [10] may thus come to be realised literally.

The consequences of extra-terrestrial tyranny are, therefore, of the most profound scope. Examining the conditions for its emergence and persistence beyond the Earth is not merely an intellectual enquiry of interest to political philosophers and scientists, nor are its consequences necessarily confined to those unfortunate enough to find themselves under the aegis of its practitioners. A successful diminution of the conditions that allow extra-terrestrial despotism to take hold may ultimately turn out to be essential for ensuring the continuity of human freedom itself.

3. What is Extra-Terrestrial Tyranny?

Any examination of the nature of despotism beyond the Earth must offer some definition of the term 'tyranny' itself. I have taken the meaning to be rather broad, involving things that invade upon the social freedoms of individuals [6]. It is worthwhile to define a little more accurately what tyranny might be in an extra-terrestrial environment.

It is no trivial task to provide a coherent working definition of a word to describe the social state of a population that does not yet exist and for societies in environments where human beings have never permanently lived. There is also a certain degree of presumption involved in even attempting to provide this definition for people who will develop their own more convincing definitions, derived from the very experiences they have of government and survival in their specific environments [11].

Another approach to this problem is not to try to formulate an accurate and inviolable definition, but instead simply to create a definition of tyranny for the sake of an enquiry into interplanetary despotism – a convenient working conception. If there are forms of tyranny that settlers eventually encounter of which we cannot conceive, then this analysis will obviously be wanting in these areas and they will have to be addressed later. If there are aspects of behaviour that we consider tyrannical, but they consider necessary, then these facets of any analysis of liberty can be ignored.

Before I yield to an expedient definition, it is worth saying more about the particular challenges presented by an attempt to define tyranny. There are two, not entirely separated, considerations in attempting to clarify extra-terrestrial tyranny. The first is the metaphysical question of what liberty means to people in extra-terrestrial environments and what normative criteria they use to define freedom. This discussion amounts to the same debate that has entranced and occupied political philosophers since antiquity. It is an argument about the meaning of freedom, the existence of negative or positive versions of freedom and the different types of liberalism possible in extra-terrestrial environments. In this discussion I will not make any predictions about how social scientists in these environments might see this discussion from their vantage point. However, I will make the assumption that societies that pursue the tenets of liberty will seek a maximum sphere of social space in which people can have private lives, but at the same time they should seek to allow individuals wide involvement in the polity, both to pursue common objectives with other people and to pursue their own idiosyncratic ideas and projects. This apparently simplistic view of a realm of liberty provides a point from which to consider the challenges to these objectives that might emerge in an alien environment; they offer a working position that can be modified, even ignored, when the reality of extra-terrestrial tyranny and its particular forms emerge within the views of liberty that settlers eventually have.

The second complication is the problem of identifying the practical and policy causes of tyranny. There is little value in drawing up a long descriptive list of all those things that the inhabitants of future space settlements might find tyrannical. Many of the matters that occupy the minds of civil liberty campaigners on the Earth today, such as excessive state surveillance, data collection, identity cards, and so forth, are matters linked to specific technological capabilities and historical contingencies. Although the generalities of these incursions, for example excessive personal data collection, are equally likely to become problems in the extra-terrestrial environment, their specific nature will elude us. When and where it is useful, I will discuss particular situations as examples of potential tyrannies, but in attempting to define generalities, this is not a useful exercise.

The character of tyranny will be dominated by the effect of the lethal environment and the social and political doctrines that derive from extreme confinement within those environments. Never before, anywhere on the Earth, have people lived permanently in places where the atmosphere around them is instantaneously lethal, or where in fact there is none at all.

However, extreme conditions or physical constraints have been a favoured example for political philosophers of the assertion that a physical inability to do something, such as to fly, cannot be considered a restriction of liberty. They have, in the past, dismissed the idea that physical extremes are a concern in discussion on liberty. Famously, Locke said 'and that ill deserves the name of confinement which hedges us in only from bogs and precipices' [12]. Mill observed, 'If either a public officer or anyone else saw a person attempting to cross a bridge which had been ascertained to be unsafe....they might seize him and turn him back, without any real infringement of his liberty... Nevertheless, when there is not a certainty...no-one but the person himself can judge of the sufficiency of the motive' [13].

Mill acknowledges the possibility of a lack of certainty in these problems and the requirement for individual judgement in determining when an external interference to protect one from an apparently unsafe situation is an infringement on liberty. But it is this matter of degree that makes the legitimacy of interference so difficult to divine. If an authority was to prevent someone from accidently opening an airlock onto the lunar surface – perhaps they are temporarily mentally incapacitated for some reason – then it seems obvious that their liberty has not been infringed, consistent with the views of Locke and Mill. But matters will rarely be this simple.

One can illustrate this with a simple hypothetical example. An extra-terrestrial state or governing body might have reason to install a surveillance camera in an area of a habitat in which there is a structure harbouring a risk of mechanical failure. The lack of its continuous observation might threaten depressurisation and death to those who happen to be near this structure. Those in command are sure that if anything does go wrong with the structure, there will be initial failures that can be adequately picked up with a camera, allowing sufficient time for a repair to be implemented. Does anyone in the settlement object to this course of action? It is doubtful that they would. Most persons would see this action as something akin to preventing a person wrongfully opening an airlock; this interference is a matter of public safety.

In the next stage, officials expand the network of cameras to other similar structures with comparable potential design faults. Prudence would suggest that they too be scrutinised. To reduce the workload, and to increase the chances of the detection of an engineering failure, these cameras are to be linked into a network overseen by a central authority. After all, since they are all monitoring the same type of structures, it seems sensible that specialists trained to spot the defects are assembled in one place and can easily communicate the information.

The imperfection that has been observed in this particular structure, it now turns out, can also manifest in a range of other engineering constructions. All of them are very different, but they use similar materials and there is some concern that these other units might eventually be at fault. A much larger network of cameras linked to the central observational facility must now be assembled. Are the inhabitants still happy? Maybe they are. But at some point there are people amongst them who must become concerned about the necessity of the surveillance and whether the benefits gained in safety are greater than the losses in privacy that these cameras cause [14].

The scenario described could be applied to any other social policy or action taken by a central authority. The specifics are not important. The example is given to underline the more general point that it is true that preventing someone from dropping off a precipice or into a bog, as Locke would have it, in extreme cases cannot be considered to be an infringement of a person's liberty. However, the true necessity of social policies that derive from keeping individuals off the cliff edge or away from the bog can be obscure and it is in these realms that gratuitous tyranny and coercion find an open field. If the choice is between instant death or giving officials the benefit of the doubt, the cautious will be inclined to allow the implementation of more invasive social policies and dictates. In more tyrannous states, the precipice, the bog or the lethal lunar environment become a convenient excuse for extending the reach of state control and reducing the number of activities in which people find themselves free of duress.

From this viewpoint it is possible to derive a working view of tyranny in the extra-terrestrial environment, and this needs to be more definite than 'a name expressive of everything which can vitiate and degrade human nature' [15]. I do not think that the description of it need be very different from that which one might come up with in the terrestrial case. Nevertheless, whatever we decide, it must successfully circumscribe the problem of the encroachment of the state into freedom under the guise of protecting the populace from the deadly environment.

One might say that extra-terrestrial tyranny is a set of circumstances where people's private sphere of activity or their opportunity to express their individual potential is gratuitously curtailed by an authority in a way that the consensus would agree, if the public were in full possession of the relevant information, was unnecessary for the wellbeing of the individual or society.

It is essential to capture tyranny as both the restriction of one's private sphere of activities (traditionally called negative liberty) and the restriction of one's potential to go out and do things that enable one to achieve certain ambitions (traditionally called positive liberty) [16]. I have delimited my view of tyranny to a situation that the consensus would agree with. Without this proviso, any individual's views become an acceptable definition of tyranny and of course that gets us nowhere; so there must be some common agreement that the incursion is wrong by a consensus view. Of course, the consensus can still be in error and there is never any guarantee that the majority of the public have developed a sound understanding of the matter. For that reason, I have finally added the caveat that they must be in full possession of the relevant information.

This last caveat is an important one. In the hypothetical scenario of the surveillance cameras, the authorities can implement extensive systems of monitoring by convincing people that their safety is at stake. All they have to do is to exaggerate the scale and ubiquity of the potential engineering flaws that they seek to monitor. The state will always try to extend its reach, out of a desire to expand its power base, but also out of a quite benevolent desire to avoid making incorrect judgements that put lives at risk. Even the most benign administration will err on the side of caution and take measures that the consensus view might decide were unnecessary. However, if individuals have full access to the relevant engineering information, the consensus view might be that the monitoring is excessive and unwarranted, and that infrequent checks by engineers on the ground are quite sufficient.

Within tyranny lies a certain degree of gratuitousness, an expansion of power for its own sake, or out of excessive caution that does nothing useful – and is often detrimental – for the people whom it purports to serve. This gratuitousness can be revealed by the test of whether the public would abide by it if they were given the opportunity to derive a consensus opinion through full access to information about the reasons for its implementation.

The view of tyranny I have provided is not fool proof. For a start, it assumes that states wishing to avoid tyranny would think about everything they did within the context of people having all relevant information available to them and thus being able to form a consensus. If extra-terrestrial states were inclined to do this with every decision they took, there would be no concern about tyranny in the first place. They are not likely to carry out this analysis well because they will make their judgements against the backdrop of unwarranted caution or power-mongering, which will lead them to a skewed assessment of what the consensus would want. Nevertheless, the view I have expressed is useful for understanding what types of cultures are more likely to result in despotic leadership and tyrannous social environments.

Tyranny is much more than direct action. The mere threat of coercive laws and regulations or, worse still, the conditions that allow authorities to wield unpredictable arbitrary power, are sufficient to create a culture of fear [18]. This apprehension can gratuitously curtail a person's ability to express or defend their freedoms. The presence of these dangers and the culture of trepidation that arises from the possibility of unexpected state attention is equally a component of extra-terrestrial tyranny.

But despotism can go further and I must agree with Mill's observation [19] that disapprobation by a majority viewpoint can also be a powerful restriction on liberty. For small groups within extreme environments, in particular, collective disapproval can be an overwhelmingly strong disincentive for individualism. For those on the receiving end, collective disapproval can be one of the most painful, cramping and efficacious forms of enforced mental constriction. The view I have discussed here, which includes the notion that such behaviour is that which is seen to be tyrannous by a consensus, would seem to allow for group coercion over an individual because the consensus might be that it is not authoritarian to curtail or crush a particular dissenting individual's liberty. Although one can argue that one person's errant views are at odds with the wellbeing of society, in most cases, particularly when it is not against laws prohibiting serious crimes, it is not the case that the expression of individual ideas and views threatens the felicity of society: rather it would seem that gratuitous imposition of a majority view on a person who has the gall to be different in their views constitutes a form of tyranny.

I think that it is important that this Millian view is accepted, because if an extra-terrestrial society does not accept that public and majority views can be tyrannical, then all manner of mechanisms, including the blatant manipulation of media and other information, can be used to nurture consensus views that seek to create a conforming, herd-like mass of individuals. A consensus view that should disapprove of some acts can be ruthlessly and dogmatically manipulated to ensure that it does agree with state edicts [20]. It is obviously the case that for a population to be in a state of mind where it can objectively and sensibly assess potentially oppressive motives and policies, it must be able to partake in a free and open discussion in all forms of media and to gather as much information as it can from different sources to derive its judgements [21]. So, an open and vigorous media is essential for the success of an extra-terrestrial settlement.

The foregoing discussion may be somewhat frustrating. I have provided neither an all-embracing and precise definition of tyranny, nor a coherent list of policies, state actions and regulations that might be regarded as autocratic. This is the case partly because I think it is not possible to do so, given the relatively unknown social conditions in any location in space and the unknown attitudes of those that reside in it, and partly because too precise a definition is counterproductive. The more one attempts to circumscribe an exact formulation of the meaning of tyranny in outer space, the more likely it is to be at odds with the eventual reality. A broad definition has some chance of overlapping with, and so providing something useful toward, a comprehension of the eventual conditions for dictatorship.

In trying to understand *a priori* what challenges liberty faces and what general direction of society interplanetary settlers *qua* liberty-seekers should try to take, I have averred, therefore, that it is possible to derive a useable understanding of tyranny. In essence it amounts to a gratuitous attack on people's private lives and on their public capacities [22] that serves no purpose other than the advance of power for its own sake, on behalf of individuals or organisations that exert influence over the social order. Tyrannous behaviour is usually self-evidently antithetical to a purposeful desire and attempt to increase the realms of human freedoms in private and public matters and to allow citizens the opportunity to express their full potential within a sufficiently cohesive societal structure. From such a general description, it then becomes possible to think about the things that will encourage tyranny.

4. The Historical Origins of Tyranny

An extra-terrestrial society does not just emerge from the vacuum, but, like any new settlement, it carries with it the historical legacy of its origins. Even the most isolated outpost in the farthest reaches of outer space must still have been placed there by some state or corporation. It is these indissoluble links with their origins that provide the historical opportunities for the appearance of tyranny.

A new society of any kind faces the question of the extent to which it attempts to independently command its own destiny and the degree to which it maintains links to the organisations that gave rise to it. Very rarely do these perceptions coincide: the founding organisations have put vast resources into establishing the colony and ensuring its success, so they see this effort as meritorious of considerable latitude to control the direction of the group. The people within the settlement are the ones who have confronted life-and-death situations and have overcome failures to ensure the viability of their enterprise. They see it as their inalienable right to determine entirely their own future. This schism between these two groups has been the problem of new colonies on the Earth since time immemorial and there is no reason to believe that things will be any different in space.

Any imposition by the founding organisations that appears

to be unnecessary or draconian will be viewed as tyrannous by the community, particularly where these dictates interfere with decisions that they perceive to be crucial to their existence. The extreme conditions in outer space, and the potentially lethal consequences of errant decisions, will magnify the friction between the founding organisation and the settlements on matters to do with the survival and organisation of people.

The roots of this potential conflict are clear to see. At the time of writing, before a permanent society beyond the Earth has even been established, the physical regions of space are subject to the United Nations Moon and Outer Space Treaties, Committee on Space Research policies on planetary protection, and a range of other measures or proposed measures to regulate the environments of outer space [23]. Will future settlers agree with these regulations? It is evident that as policies on the Earth concerning the treatment of places in space grows, so the future potential for conflict with those who are sent to settle in these locations will expand.

There is little that can be said about these conflicts, as their resolution will depend upon the specific circumstances of each one. However, history is on the side of the settlers who, being the people faced with the realities of the environment in which they live, must fashion their own society and determine the liberties that are important to their continued existence. It is they who must take ownership of the land and develop it in any way that they please, within their sphere of influence, if they are not to be subject to incoherent decisions from external organisations.

Historical origins of tyranny in extra-terrestrial settlements will therefore emerge from organisations, the function of which does not require them to pursue liberty or, in more extreme cases, from states that are altogether illiberal. At the time of writing, all major space exploration activity is carried out by state organisations of one kind or another. These organisations do not pursue deliberately autocractic policies, but their culture is one of state direction of space activity, overseen by national governments. If any of the organisations were to direct the formation of a commune on another planetary body, the culture would be one of a highly directed and planned mission, with specific objectives and milestones to be accomplished. The tendency toward this mentality does not reflect an inherent desire to control, but merely the pressure to deliver on taxpayer's money and to prevent a disaster, for which the public would hold them responsible. None of these organisations is charged by their public to pursue the tenets of liberty in any particular way, and it is not surprising, therefore, that no overt attempt to incorporate the precepts of liberty into a new settlement would be expected from them.

The tendency of these organisations to establish control cultures is fostered by the sheer scope of resources and the marshalling of human capacities required to launch endeavours of any significant scope in space. As these enterprises become more ambitious, so the tendency to centralise, coordinate and control when confronted with the resource and human requirements of space exploration and settlement is furthered, leading to more powerful, bureaucratic organisations. There is no reason why these organisations should be especially against liberty [24], but single organisations charged with pursuing unitary agendas throughout the immensity of outer space are likely to stifle dissent and imbue space endeavours with highly invariant structures just because they must do so if they are to achieve their goals efficiently. These characteristics are amplified by the arrogation of power by individuals in these large structures and the opportunities that expansive networks of administration provide for groups of individuals to develop specific and unified visions of space settlements.

Without liberty considered in the planning, implementation and development of a space settlement, several difficulties will result. The scenario elaborated above, whereby the settlers begin to feel tyrannised in their activities as they seek greater freedom to decide their own affairs, is more likely to be realised. The result of oversight by large organisations will be a naturally hierarchical chain of command that reflects the attitudes and culture under which the society was established.

I have previously made the point that the physical architecture of outposts can be engineered in such a way as to maximise the chances for liberty to flourish [25]. However, without the compulsion to design the physical spaces of a settlement with liberty in mind, these opportunities to advance the state of freedom will be easily lost.

When these influences are combined, the first settlements, particularly those established by government organisations, are likely to drift into a more unyielding, rigid configuration with less emphasis on individual liberty, and this legacy will leave in place the institutional constitution that will tend to sway them toward social structures that are more conducive to servitude. Societies established by military organisations will approach this reality even more exactly, because they are more likely to enforce strict systems of hierarchy and discipline than civilian government organisations.

It is obviously also the case that this state of affairs will gain even greater force when the founding organisations are from nations that are illiberal and seek to use the space frontier to advance fundamentally totalitarian doctrines of social organisation. Communities arising from such a scenario will have imposed on them the double tyranny of illiberal ideas inherited from their founding states or organisations, and the repressive influence of regimented regulations to protect against the extra-terrestrial environments.

The foregoing might suggest that the best way to achieve liberty is, therefore, to encourage the establishment of settlements by private organisations. Insofar as they are likely to pursue a more commercially driven agenda, which may, through competition with other operators, encourage more liberal conditions to satisfy customers, then they may be successful at establishing the conditions necessary for liberty to flourish. But private organisations are not immune from the potential to give birth to tyranny-prone cultures. Large corporations, possessed of vast resources, can be just as ruthless and calculating as state organisations; particularly if they should find themselves in control of local monopolies of resources or other profit motives.

And it must be remembered that tourists and other types of individuals who live in space may themselves demand high levels of central oversight to protect themselves from the environment; tyrannous edicts may be demanded from the very people for whom liberty is being secured. Human history is littered with countless examples of despots supported by, or brought into being by, the acquiescence and quietism of the very public subjected to them because they secure a measure of security, common direction and predictability.

Mitigating the chances for despotism to emerge from the early relationship of settlements to their founding organisations can be achieved, but it should be remembered that these are only the beginning of the settlers' trials. Ensuring that the loss of oversight from the originating organisations does not leave a settlement at the mercy of its own, even worse tyrants, will present a challenge of equal proportions. From where can these local sources of tyranny emerge?

5. The Environmental Origins of Tyranny

Survival in a lethal environment, which is the challenge of existence for all interplanetary settlements, is a matter of life and death for the very continuity of the population and the society itself. In many ways, it can be viewed as the moral equivalent of war. The common enemy against which a people must struggle to survive is well-defined, it is pervasive, it is powerful and it threatens to destroy those who do not successfully mobilise against it. It will be trivial for an extra-terrestrial authority to justify procrustean measures in order to range the resources of the settlement against this common foe and to ensure the survival of all members [26]. It is this simple reality that provides the most thorough-going channel through which tyranny can emerge.

The extra-terrestrial environment's tendency to solidify and give succour to tyranny works at many levels. Its most powerful influence materialises both from the extremity of extra-terrestrial conditions and the spatial isolation caused by the vast scales of outer space. Until a very large number of outposts are established, the option to emigrate from a settlement is either non-existent or difficult to realise and it is this imprisonment that provides authorities with a degree of latitude in their control over people which rarely exists on the Earth - except in countries that have successfully secured their borders to dissenters or amongst the impecunious who do not have the resources to move.

In terms of the influx of new ideas and opinions, extra-terrestrial environments represent something of a throwback to medieval conditions. One of the great contributions to liberty during the twentieth century was the technological weakening of borders between terrestrial nations. First, radio waves penetrated these dividing lines and then the internet, which provided the population of even the most draconian of states with outlooks and views on their own situation that allowed them to assess more objectively their own predicament. Even if the information is, in practice, useless, in that inhabitants are so confined by physical borders that they cannot escape, it keeps alive a flame of freedom in the mind, a comparative perspective from which they can take an unsullied view of their world.

Spatial isolation makes it difficult for other organisations from afar to transmit information to a tyrannised population. Even if there is no deliberate repressive intention on the part of their authorities, there will be a natural tendency for settlements to become insular with respect to the information flowing into them. Deliberate attempts to beam information of various kinds will be frustrated by the spatial separations between outposts and the delay in transmissions caused by the finite speed of light, which will obviate the possibility of instantaneous communication. Outer space is a barrier to the free exchange of ideas – a barrier imposed by both the vastness of the territory and the laws of physics.

The combined effects of both the physically extreme environment and the culture it propagates, and the spatial isolation of people with respect to movement and new information, contribute quite simply to the culture of conformity. Indeed, I would hazard a guess that the word 'conformity' will become one of the most important words in the lexicon of interplanetary settlers concerned for the conditions for liberty. Everything about the extra-terrestrial environment converges on this word. Not merely the conditions I have just summarised, but the utterly monotonous colours, sounds and smells that will be associated with the daily tasks of life in space or on other planetary bodies, the lack of a biota to add variety to the landscape and infuse the senses with the luxuriant spectacle of the products of biological evolution, and the narrow range of industries and leisure activities that people can turn their hands to will, at least in most of the early stages, all contribute to conformity. The intelligence of the population will be no guarantee against this trend [27]. Organisational structures within settlements will intensify conformity. In those with a small population size, there are good reasons for citizens to combine into organisations and pool their efforts [28] so that even in environments where voluntary association is encouraged as the dominant means of social organisation, conformity will be rigidified.

An essential precondition for a politically and economically open state, and the minimisation of the convergence on conformity, is the ability for individuals to gather and associate free of interference. It may well be one of the most interesting paradoxes that often these places to assemble and dissent are public spaces owned by the state. Although one obvious answer to this problem is that the politically-minded can convene in private habitats, there is something quite conducive to a revolutionary fervour if people are induced into secretive meetings behind closed doors. If public engagement in the polity is to be secured in an open and positive way that does not produce the sense that dissent must be done *sub rosa*, away from the prying eyes of the state, then a more healthy democratic culture will flourish. On the Earth public places are high streets, parks and similar open areas. In the extra-terrestrial case these places are pressurised walkways, common modules etc.

This problem of conformity is important to understand because one could go on, seeking every avenue of possible tyranny, codifying and quantifying the sources of despotism and attempting to uncover every recondite threat of coercion that results from the environment and the culture it spawns. If all of this is reducible to the common problem of intense conformity, then we have identified the key threat to extra-terrestrial liberty. And if it is the case that this is the central problem, then an authority does not need to know about, or need to seek to identify, every possible policy that will lead to tyranny. All that is needed in order to achieve a culture of liberty is to pursue general policies and encourage ideas that minimise conformity and maximise the plurality and expression of disparate opinions in a settlement [29]. If this is achieved, then the main enemy of freedom can be slain and other liberty-seeking policies will be added veneers of freedom on this fundamental structure of liberty.

The endpoint to which extreme conformity pushes a society is to a population of contented extra-terrestrial slaves [6] – a cryptic natural tyranny in which the authorities become slave owners, and the vassals, constituting most of the populace, are oblivious to any state of social organisation that could improve their lot, spending their time nurturing their fealty to their extra-terrestrial lords. Like any scullion, their sense of liberty rests on the fact that they do not recognise the potential they have. Confined to a small range of tasks, jobs in which they perceive themselves to be free and provided with the anodyne security of their masters, they lack any understanding of what liberty they might have in an environment where conformity was not so rife [30].

If the slaves are a happy lot then one might argue that they are all free in a very egalitarian sense and this vision of liberty, in which the population declares itself to be free, might seem conveniently justifiable to authorities. In as much as these people are well-behaved and satisfied, it would be easy for such a society to be defended by those who run it. How much more dangerous is a society of the inquisitive and acquisitive, who are constantly seeking to expand their potential and their opportunities for action in an inherently lethal environment? As any of these actions are unpredictable then it is also easy to understand why a society which sees liberty as amounting to the duty to expand people's opportunities for action might seem to be a potentially precarious and actually irresponsible type of society. The contented extra-terrestrial slaves may be in chains, but better that they are in chains and happy than asphyxiated and dead.

The presence of this pressure lays bare a schism in concepts of liberty that will be particularly acute in the space environment – the difference between negative and positive concepts of liberty. The negative idea of liberty focuses on the lack of restraint from doing things, a sphere of individual activity that is outside of state coercion or other forms of control. This is the conception of liberty invoked by Hobbes, Constant, Bentham, Mill, Hayek and others. The positive sense of liberty is about the power and opportunity to do certain things. In the sense that freedom is the lack of restraint curtailing the population's activities, then it can be argued that the negative and positive senses of liberty are just a focus on the two different parts of this relationship [17] and so their separation is logically specious. However the separation is particularly useful here to emphasise the problems of liberty and conformity in an extra-terrestrial environment within the general relationship of freedom.

A focus on negative liberty by an authority is one way in which to reduce those restraints of social practice that invade people's lives. Accepting that no social policy can relieve the reality of the instantaneously lethal environment, the social policies that result from the influence of the environment can be engineered in such a way as to achieve safety without excessive control of lives. However, negative liberty cannot abrogate the problem of the contended slaves, who, if they are happy in their highly circumscribed lives, will accept a far smaller range of liberties without dissent in the first place. If, through the information they are fed, their upbringing or the conformity that has taken hold in a society, they expect few freedoms, then very few policies will be considered to be an invasion on their lives. Worst still, the authorities can hold up a series of impressive objectives to reduce state interference, giving the impression of a highly libertarian society, when in fact the mechanisms for control are pervasive, but their effect is completely unseen because the inhabitants are naturally happy with their set of minimum freedoms. In some extraordinary circumstances, the community may even grow to share an unremitting, numinous love for their authorities for securing them their banal freedom against the lethality of the environment [31].

It would appear that the pursuit of negative freedom is, in itself, insufficient. In a society where conformity is the pivotal problem caused by numerous facets of the extreme environment, then some type of pursuit of positive liberty is necessary [32]. By this, I mean policies that enable individuals to enlarge their potential for action and enlarge the potential for those opportunities to be realised [33]. The practical means by which this is to be achieved may include education, the opportunity and encouragement to be engaged in political and economic activity [34], and transparency in the political and economic means of production and supply (as discussed in [5]).

This aspect of positive liberty should be very succinctly separated from the type of positive liberty that emphasises the liberation of the individual through their opportunity to engage in the wider political and economic process in activities that are commonly agreed by everyone, rather than more individual activities which are entirely a product of their own personal creative ideas. Understanding the distinction here is especially important for the extra-terrestrial case because it will most certainly be true that the liberty of an individual to be able to express ideas and plans will be realised partly through the combined and common actions of many individuals.

There is a plethora of examples of cases in which this second type of positive liberty will be realised. The construction of habitats and new living space in the extremities of the interplanetary environment is very difficult for an individual to achieve alone; in fact it is virtually impossible for an individual to do anything useful with extra-terrestrial land without being part of, or mobilising, collective effort. The security that allows individuals to express their own views and creativities depends entirely on commonly agreed safety standards and the collective oversight of habitats and other infrastructure on which survival depends. The lethality of the extra-terrestrial environments creates powerful and unique pressures on a collective that force it to work together [35]. Thus, the liberty of the individual is inextricably linked to membership within a greater whole – a society. This view of liberty is very much in tune with the views of Burke and other conservatives [36], and it sometimes engenders within it a belief in progressive change as the means to social advancement, rather than revolutionary transformation [37].

Within this positive sense of liberty, there is a similarity to the ancient Greek view of liberty as representing the freedom to partake in the polity, an awareness of civic duty which allows the individual to achieve their full potential in the city-state, without which they would be a shrivelled and politically powerless creature. In the extremity of space the individual is also powerless and it is through a contribution to the settlement that they achieve the ability to be an important and influential part of a social whole [38].

Positive liberty was very much at the core of Rousseau's vision of a part of the human mentality that is in coincidence in the community and which constitutes the part of will that he termed the 'general will'; distinctive from one's individual aspirations. Hence emerges his conclusion that the deviant can be 'forced to be free' through the obligation to correct their behaviour such that their general will is in coincidence with others. One does not have to go as far as Rousseau's conclusion that the independently-minded who do not agree with the collective should always be forced to be free. Instead, one can accept that for every social decision there will be dissenters and leave it at that. The acceptance of this difference of opinion does not constitute weakness or the tolerance of a failed general will; it is just a healthy difference of opinion that should be regarded as a sign that a plurality of opinion is still very much alive.

It is in the realm of encouraging a diversity of views and ideas that these two senses of positive liberty must be balanced [39]. The first form of positive liberty, and the form that concerned me in the prevention of the populace becoming contended slaves, is the positive liberty that does not focus on the common weal and common social ideas, but it is the positive liberty that encourages people to express ideas that are not in coincidence with those of the rest of society, those irritating expressions of individualism that will annoy the authorities, but which will allow individuals to be different [40]. These personal expressions of creativity and choice contribute to the need that individuals have to express themselves and their own idiosyncrasies. However, they are no less important for generating the plurality of ideas upon which society itself depends for its flexibility, its capacity to adapt to new circumstances and the source of creativity needed to prevent itself from becoming a homogeneous mass of domesticated slaves.

There is no great difficulty in understanding why this balance will be so difficult to maintain. In one sense both of these types of positive liberty are in agreement, for if individuals express their own creative powers, they will be contributing to the diversity of ideas in the polity. On the other hand, if these more personal ideas should be out of synch with collectively agreed ideas and norms of civic duty, which individuals also use to express their powers and liberate themselves within the culture of the collective, then the positive liberty that espouses the expression of individual ideas may fall prey to the positive liberty that emphasises the importance of conforming to the collective will as the route to personal involvement in the polity. This can occur because the latter is espoused by a great majority, whereas the former is, by definition, the expression of a single individual. The crucial point is that the very arguments for positive liberty can be used to crush individual freedom. Although common civic duties will be a required part of a successful society, they should not be confused with the encouragement of self-expression and the self-realisation of individual potentials – nor need they be in conflict with it [41].

Attempting to quantify conformity in society, which erodes these senses of positive liberty, let alone being able to decide what is too little or too much orthodoxy, is an interesting, but essentially valueless task. One would tend to allow more individualism in activities that did not potentially threaten lives, such as what type of food would be grown. One would probably allow for less individualism in a matter with potential for more drastic social implications, such as policies influencing oxygen-producing entities. Any attempt to quantify conformity, then, may well be counter-productive by generalising on a matter which must be decided in different particular cases.

Despite this, it is possible to realise that reducing conformity, when and where it is possible, must be a guiding principle of a society beyond the Earth. If such is the case, then many of the threats to liberty that have been discussed here can be overcome. Encouraging new initiatives and views will reduce the uniformity of central dictates, lessen the chances that negative liberty will be eroded, and it will encourage a general culture that sees the power of individuals to engage with the polity using their own ideas, aspirations and ambitions, as a bastion of positive liberty to reduce the chances of civic duty and collective peer-pressure transforming into instruments of tyranny.

6. The Causes of Economic Tyranny

Causes of political tyranny are not obscure. Systems of control and the culture of coercion that result from the extremity of the extra-terrestrial environment, and the need to counter them, are the major components of an erosion of freedoms. Nonetheless, equally powerful forces in this direction are likely to be found in the economic realm; their origin is much the same as the origin of political dictatorship, but their manifestation and the mechanisms by which they might work are likely to be very different.

In the economic arena, far more than in the political realm, it is isolation that will bring about the conditions for economic tyranny. In the political sphere the local environment is the most prominent cause of despotism because its effects are so present in people's daily lives. The extreme isolation contributes to this core problem by reducing the ease of emigration and the exchange of information. In the matter of economics, isolation will define the economic character of a settlement, manifested particularly through its opportunities to trade.

Economic tyranny can be most clearly manifested in the acquisition and processing of raw materials. For any type of entrepreneurship to succeed it is vital that a business not merely gets access to the raw materials it needs to make its product, but also that it acquires these predictably over a long period of time in order to fulfil customers' demands reliably. For most products, these resources will be difficult to come by locally partly because the extreme environments of outer space make it difficult to go and collect resources without considerable logistical networks run by others, and partly because, in all likelihood, not all of the required resources can be acquired locally anyway (by locally here I mean on the same planetary body). The latter is particularly the case on small asteroids and in the gulf of free space.

These factors bring into play two potential sources of despotism.

First, if resources are more difficult to acquire, this leaves the simple matter that occasions when these resources cannot be acquired (for example in the breakdown of delivery spacecraft) are more likely to crop up. Thus, there is the problem of the sheer challenge of working in the space environment that exerts itself on any organisation trying to carry out dependable, predictable business. This is hard to classify as true tyranny, as there is no human agent involved; it is a simple reality of the nature of operations in the space environment, but its effects may well be felt through political channels if the inadequate supply generates a culture of control.

Second, the complex logistical exercise in acquiring, extracting and transporting raw materials across the Solar Systems, and the likely involvement of large organisations that command space transportation networks, reveals the possibility for genuine economic tyranny in these supply and production links. Any of the connections in these supply chains are open to interference and coercive abuse by the organisations involved in them or, through regulations and safety demands from authorities either wishing to influence these organisations directly or to prevent them from providing resources to a business entity causing them nuisance within their own political sphere of influence. The problem amounts to the uncontroversial observation that the greater the number of links in a supply chain and the greater their complexity, the greater the number of opportunities there are for intervention in their successful implementation.

The sheer physical difficulty of extracting any resources in space and transporting them makes it hard for a business to bypass other organisations and acquire the materials for itself. Competition can help circumvent the problem. It will not mitigate the possibility of an authority using legislation and other regulations to control a business or industry, but it would solve to some degree the problem of the need to ensure choice, and thus reduce monopolisation, in the provision of raw materials and other processed products to space industries.

One could imagine a picture where these controls were minimised. Consider, for example, a situation where economics was run entirely on a market model with no oversight from a governing body and the implementation of a wholesale *laissez-faire* economy was the order of the day. There are reasons to believe that this would be impossible from a number of perspectives. In the lethal conditions of space, someone must ensure that processes for the production of vital commodities for survival, such as food growth, water acquisition and oxygen production are secure. Now, whereas one could argue that because these are so vital to human life, there would be immense market demands to ensure they are provided and that their sheer essentialness will ensure the success of market forces, there can also be little doubt in saying that as on the Earth there may be market failures such as the closure of businesses selling vital space parts and so on and so forth. As these types of occurrences could potentially cause a catastrophe, or the loss of the required quality of the product, there will need to be an authority, even in the perfect *laissez-faire* economy, able to step in to run a business or secure a vital commodity or set the safety standards to ensure that products are being made to the required specification. The emergence of a type of mixed economy in this way, or at least some minor form of state oversight, need not lead to any particularly tyrannous state of economic affairs if it is judiciously managed.

There are other inducements to increase control of the economic sphere and they revolve around the problem of employment. An authority cannot allow large numbers of unemployed. The lethal outside environment and the threat they would pose if they turned to criminal behaviour, particularly destructive criminal activity, means that they must be kept occupied. As vagrants and the indigent within enclosed pressurised spaces, particularly those with a grievance against the authorities, represent a grave threat, there will be immense pressure to find employment for everyone. Even if they keep themselves to themselves, the unemployed will still be an economic and social burden with little useful economic contribution, so they are a wasted resource. Thus, extreme conditions will ineluctably force upon a governing body a policy of full employment.

Now some economists would raise an eyebrow at the claim that full employment represents a form of tyranny, especially those of a Keynesian persuasion. In a confined space this policy is not so much economic tyranny in itself, but it offers an excuse for gratuitous control. With nowhere else to go, the employees could find themselves at the receiving end of a policy of full employment that moves them from this and that industry and tells them that they must do this and that job [42]. The apparently benevolent need to ensure that everyone has a job can be used against specific dissenters and troublemakers to remind them of who is in control of their livelihoods. In addition to this, by pursuing such a policy the authorities can exert power over corporations, forcing them to take on particular persons (some of whom might even be state functionaries), in order to fulfil their obligations to the greater social good of preventing unemployment.

It will be difficult to move unemployed people to new settlements, unless there are locations close by. If Solar System-wide transportation networks eventually become available, then personnel can be moved amongst companies in much the same way that employees move between jobs on the Earth, avoiding much of the pressure to achieve an artificially directed full employment. One should add that the use of prison in extra-terrestrial settlements and the usual systems of criminal prosecution can be used to deter the unemployed, and indeed any citizen, from carrying out destructive actions. However, despite these fallback positions, the point I want to make is not that policies of full employment are absolutely required, but rather that the excuses to pursue such ideals, and the small population sizes that make them look viable, provide a mechanism for exerting economic tyranny that can be used by authorities or even private corporations seeking to advance their power base.

Full employment will never be completely achievable even if it is pursued by the extra-terrestrial state. There will be people in transition between jobs, people who cannot find work and inevitably the unemployable. The lack of the multiplicity of industries that are associated with the billions of people on the Earth, and the diversity even within some of the smallest nation-states, will increase the probability that there is no job to suit someone's specific skill set, or no available work in a specific type of activity at any given point in time. All of these factors may conspire to make some type of Welfare State necessary -

some type of unemployment benefit or insurance. By extension, the provision of a Welfare State will require central economic resources and eventually, in the longer term, when an outpost becomes large enough, a banking system and currency to make all of these transactions possible [43].

Related to this and linked to it is the effect that interplanetary environments have on the division of labour. In their early stages there will be a strong incentive to select those people with general skill sets, or those who are trained to be able to work many different types of machines and processes. The incentive for this emerges from the need to ensure redundancy in the population and that no single and vital skill is lacking. As the population expands, this pressure is reduced. However, even in quite sizeable populations there may still be pressure to train people with multiple skills so as to reduce the risk of unemployment; indeed, individuals themselves will have an incentive to make their skills broad and widely useful.

As Bernard Mandeville and Adam Smith recognised, the division of labour is the key to high industrial productivity and efficiency, necessitating a balancing act between maintaining a population that has the redundancy required to maximise its viability and impelling the generalists to acquire specific skills that allow them to contribute to economic tasks in which an outpost can be competitive, particularly if it involves itself in external trade with other settlements, or even the Earth [44]. The desire to control this balance may tempt authorities to interfere with, and direct, employment patterns. The power of the employee to resist this interference will be further weakened by the reduced power of employees to challenge their employers [45], particularly in vital industries. So unpopular will any strike action be that it is not likely to succeed and so the threat of it is neutralised. If treated inappropriately, it will be difficult for employees to find mechanisms to change the behaviour of their employers.

All of these things to which I have alluded are important elements in the potential for economic despotism, but there can be little hesitation in saying that the tendency to autarkic economic conditions in extra-terrestrial environments is the greatest challenge. Separated by enormous spatial scales and divided by a testing environment, even from other settlements on the same planetary body, societies will suffer relative autarky compared to their terrestrial counterparts.

Autarky caused by spatial isolation will influence economic activity in many ways. Lacking the plurality of contact with other industries and particularly external competition, specific industries can easily fall prey to cartels that control production and supply, as well as the movement of labour and prices. With few external organisations to increase competition, to drive down prices and increase the healthy ebb and flow of people, economic conditions would risk stagnation and subsequent decline into a generally venal, insular and corrupt frame of mind. From this condition, the chances for monopolies or oligopolies to seize control of vital industries is increased and economic tyranny will ensue, while the level of corruption and economic manipulation will multiply [46]. The forces of economic tyranny will lurk in any product that is essential to society and that is required by people who are not able to move easily once coerced.

Autarkic conditions almost invariably cause a shortage or a need for rationing of commodities. Lacking the influx of new goods, which can be traded for a surplus of goods produced by the settlement, the range of products on offer becomes restricted and merchandise is either in glut or scarce. That which is scarce must be distributed by some means, which can be either by straightforward market mechanisms – sold to those who can afford the high prices that result from their scarcity – or if the social consequences of this are unbearable, particularly for vital products such as clothes or food, then through some type of ration system [47]. The effects of this on individuals are almost certainly negative. They become prey to these organisations and their choice of employment is reduced. Economic tyranny essentially reduces to political despotism as the range of employment choice is effectively, and possibly deliberately, reduced, resulting in a population enslaved by their employers.

Conditions such as this lend themselves to a vicious circle. The decline into political dictatorship will itself encourage corporations and authorities to constrict the flow of goods and new businesses into a settlement in order to maintain economic control over supply, which in itself improves the scope of control over people [48]. Such a curtailment of economic activity reduces the threat of competition, economic and political, that might arise from the establishment of new branches of production and supply. The results of this policy would be a further reduction in political freedom and the power of the individual to express political and economic ambitions. The 'natural' condition for autarky that arises from the vast spatial scales of outer space is one of the greatest threats to both political and economic freedom. Achieving trade routes will be essential if any semblance of liberty is to be maintained in space [49].

There is no inevitability in the negative course of events I have discussed. Much of the escape from this scenario depends upon the successful development of cheap, regular and robust transportation systems through the Solar System, which allow for trade to occur reliably and continuously in order to prevent the economic distortions that crystallise under wholly autarkic conditions. There is every reason for settlements to play an active role in encouraging this development and preventing isolated economic enclaves developing, for not only will this lead to the development of more open and less corrupt island societies, but it will also enhance the economic health of the settlements themselves. Despite enormous separation of outposts, and the fact that small isolated communities will start off in a quite economically weak state, with high costs of manufacturing, there is no reason why trade, even with the Earth, should not be profitable for both the Earth and extra-terrestrial settlements, guided by a kind of 'interplanetary invisible hand', to borrow a famed metaphor from Adam Smith. There is no reason why a traditional type of Ricardian economic exchange should not be able to occur [50], leading to a protracted, sustainable and vigorous expansion of the sphere of economic activity beyond the Earth, even when the extra-terrestrial settlements appear to be economically disadvantaged compared with the Earth.

Essential to the encouragement of economic liberty will be, as I have previously discussed in the context of environmental tyranny, the ability of people to associate free of surveillance. Some form of economic tyranny will develop if all spaces fall under the ownership of specific people and organisations and the dissent that occurs within them is always subject to their good will in accepting it. Equally, it should be said that the authorities must leave spaces they oversee unobserved so that the enquiring can discuss political and economic issues without the fear of being monitored. Perhaps paradoxically, for public spaces to exist there must be common ownership of them and this implies the existence of a state that owns these spaces independently of powerful individuals or corporations, provided mechanisms exist to prevent the state from coercively monitoring these spaces.

As with political tyranny, the interplanetary environment is friendly to economic tyranny. Manifested through the control of resources and employment, both exacerbated by physically-caused autarky, the purpose of the state must be to encourage the conditions for trade. It must develop the open exchange of information and assist productivity essential for the viability of the settlement without pursing policies of excessive interference, such as forced full employment, that will yield the conditions for despotism.

7. The Problem with Lunar Leninism and Martian Marxism

There is a matter that I wish to take up concerning a point of view that emerges both from the foregoing and general considerations of the restrictiveness and collective effort needed to survive in the space environment. If it is the case that people must work together in ways never before witnessed to survive the lethal environments in outer space; if it is the case that the resulting social pressures tend to push society towards collectivist means of economic and political organisation; and if it is the case that these challenges also impel producers to ensure with diligent exactness that everyone has access to vital resources such as oxygen, food and water, then why resist all of this evidence? Is it not the case that all of these factors not only create the conditions for a type of centrally planned socialist or even communist society, but that outer space may indeed be the place where the collectivist nirvana envisaged by Marx and Lenin can finally be realised, not only because the conditions are conducive to it, but because there is an historical inevitability about the requirement for this type of society for people to be able to survive in space?

Before challenging this view, let us be clear about some of the things that might encourage a person to believe in the Elysian promises of Leninism or Marxism in outer space. The historical backdrop for Marxian views obviously has no relevance in space.

There is no one living there permanently at the time of writing and so there is no class struggle, ultimate collapse of capitalism, corporate abuse of the proletariat or any of the perceived depraved social conditions in outer space that might motivate a nineteenth-century-conceived Marxian type of revolution. So what we are discussing here is not the motive for overthrowing an existing capitalist state, or allowing its inevitable collapse and dissipation into a stateless society. Any inducement to pursue Marxism in space, at least in new outposts, must derive from what it promises to be able to do on the socially blank template of an extra-terrestrial settlement.

Whether capitalism would first be needed to achieve a Marxian society is also not a matter to deliberate here. In view of Marx's recognition of the productive power of capitalism, it might even be that extra-terrestrial Marxists would accept capitalism with alacrity as the necessary first stage in the process of establishing a productive commune, prior to its eventual overthrow or its inexorable and inevitable demise. It can be left to Marxists to worry about these details. The concern here is with the endpoint envisaged.

Amongst a variety of possible benefits, there are several aspects of a Marxian society the potential appeal of which is easily identified. They are the possibilities of a stateless society; the exact control of the supply of commodities through a centrally planned system; the lack of private property; the equality that can be achieved through this system; and the sense of community that might result from the whole apparatus as defined above. It is worth examining each of these apparent benefits in turn and where, in an alien environment, problems to their realisation might lie.

Statelessness would seem to solve many of the problems that I have expatiated here and elsewhere concerning tyranny in space and the opportunities that exist for it. Without a central state, the amorphous Marxian society would be free of the mechanisms of central political control that might erode freedom. By coordinating all action through small committees and collectives, the overbearing state apparatus that threatens to engulf an entire settlement can be held at bay.

The problem with this vision is no different from its historical problems on the Earth, but it has particular complications in space. Contemplate a fire in a major habitat or module that contains food production units. The fire has the capacity to destroy a major food unit, so the implications of this for total food productivity must be considered and those involved in the production of sustenance, and its re-allocation to other areas or corporations that produce food, must be consulted and involved. The fire, in destroying the module, could threaten instant depressurisation, so that air supply entities and those charged with preventing catastrophic air loss must be called in to deal with the problem. And then there is the simple problem of the fire itself, a fire service must come out to quench it. Following this, repair organisations must come to make the module safe.

In this one débacle at least four different groups must be contacted and co-ordinated. In the dangerous confines of space we cannot get away from the need to have some type of authority that coordinates multiple organisations in the event of a major threat to safety and coordinates them expeditiously. Now call this what you will, a Worker's Committee, a Safety Co-op, a Settlement Collective, but at some point these committees will all need to coordinate themselves to ensure the totality of safety in an outpost. They will also require access to information about architecture, safety resources, the location of persons and commodities, the repair networks and so on. In essence, given the centrality of safety in outer space, they will become the state. Anything less would, in fact, be dangerous. No depth of utopian dreaming will escape the fact that a state of some sort will emerge; although of course we can play name games, indulge in sophist contortions and deny that it is such a thing. In the very early stages of the development of an outpost, it may be little more than some type of powerful committee. It seems that it will be more dangerous to try to deny that this apparatus is becoming a state, with the risk of sleepwalking into the development of a totalitarian culture, than it would be to establish this organisation as the state in the first place, subsequently using legislation, laws and Constitutions to control its powers.

Given some of the collective threats faced in space, particularly safety threats, is it even desirable to have a stateless society? The functions of a central body or state seem vital for co-ordinating responses to potentially disastrous fatal accidents, and although the power of the state must be controlled and curtailed to prevent the drift into tyranny, it would seem that some type of state is actually a positive benefit.

What of the central production and coordination of all commodities? The difficulty in both extracting raw materials and processing them in most locations beyond the Earth might give the impression that these efforts should be centralised. In the extra-terrestrial environment some commodities essential for human existence, particular food, oxygen and water cannot be easily acquired from the natural environment by an individual on their own. The chances that catastrophic failure would result in a paucity of these necessities would also seem to militate that supply be centralised and made predictable. Finally, the three commodities just mentioned, although the most essential to human survival, also happen to be quite quantifiable in terms of the requirement that people have for them. Taking some type of calculation of the average time at rest and in doing exercise, it is possible to calculate the total daily calorie input a person requires in their food, the number of litres of water that they should drink to sustain themselves and how much oxygen they must breathe to stay alive [51]. With a population of a given size, then, the total amount of food, water and oxygen that must be produced each day can be exactly derived and, for this reason, it is amenable to central production [52].

The same argument may be applied, although perhaps with less persuasiveness, to many other goods in a small settlement. Although other products are more at the mercy of fashions and whims, a small population with access to a well-linked computer network has a reasonable chance of feeding back demands at sales points to the means of production and achieving a responsive, centrally-planned economy which is sufficiently sensitive as to avoid reliance on a pricing mechanism as a means of controlling industrial output.

The Siren call to centrally planned production is a strong one and, in small isolated settlements, there is in fact every reason to suspect that some type of planned system of industrial management could be implemented with a certain degree of success.

In many facets of production it is very likely that settlements will start with some sort of central planning. Derived as they will be from states, space agencies or corporations, it would be very unlikely that they would begin their existence with multiple competitive entities engaged in producing each and every product required, even if they did wish to achieve this.

However, there are good reasons to reject a centrally planned economic model as a systematic, long-term desired end-point. There are economic arguments which support this view. Reliance on a central production unit seriously comprises redundancy of productive capability and, although settlements might begin from this position, everything should be done to expand the number of industries that produce the merchandise. Multiple units of production could still be operated centrally, but failure in the common management, if it affected all units, would be calamitous. These observations alone support the assembly of independent organisations operating different equipment that manufacture the same product, particularly when vital commodities such as oxygen are at stake.

Perhaps the most important economic argument is that a centrally-planned system of production is likely to become autarkic, bearing in mind the political problems outlined earlier. In principle, there is no reason why a centrally-planned system of manufacture should not trade its products with other entities, but in reality if there is no profit motive for the organisation to do so, it will not achieve the same level of inducement to put the effort into overcoming the difficult obstacles of interplanetary trade if it is only charged with fabricating enough of a product to satisfy domestic demand. The problem of autarky is just one of the many problems associated with the system, which ultimately lends itself to political tyranny.

The centrally-planned economy is unlikely, despite best efforts, to truly produce what people want and, like centrally-planned states on the Earth, it cannot predict fashions and desires in the future that will necessarily make its economic output limited and dull compared to private entities, which are constantly striving to try to sell consumers new items. The worst effect of the strictly controlled economy will be the stifling of individual creativity, the opportunity to combine to produce, and the emergence of the political culture that results from the need to generate all the functionaries and state officials, with their attendant powers, that will be required to do the planning in the first place.

The logic of a centrally-planned economy, and the attraction of this in the face of the possible failure of entities producing things so basic as oxygen, should be resisted even though it may lead to a less ordered and structured economic network [53]. The role of the state in this schemata should be to ensure that sufficient entities exist (and more so for redundancy) to produce what is needed and to encourage a vigorous growth in these industries.

Rejecting a centrally-planned economy would imply competing means of production, which itself almost certainly implies the presence of a system of private property. The public ownership of all goods might appear superficially to be a security against the possibility of people going short of vital goods. In the early stages of the establishment of settlements, it might well be the case that public ownership of certain commodities such as water and some food provisions will be required to ensure that they are distributed fairly to all occupants. However, for all the reasons just adumbrated, some incentive for production must exist independently of the people running the settlement. Quite apart from this, the problem in a highly isolated group is that complete control of all property by a single authority opens the door to political tyranny.

An attraction of a Marxian society might be the economic equality that would result from the previously discussed mechanisms. Central planning, in particular, would obviate the chances of single, private entities accumulating a vast proportion of the wealth and individuals associated with these organisations becoming their own economic tyrants. There is merit in this view, and achieving equality would certainly prevent this outcome. Yet, regulations on monopolies and other tax mechanisms could be used to some degree to prevent excessive and despotic accumulations of wealth.

Aside from the most severe cases, there are rational motives for allowing inequality. The environment of outer space is lethal, difficult to work in and a costly place in which to establish enterprises. To rely on the establishment of large networks of human settlements throughout the Solar System, solely on the back of state enterprise and centrally-directed orders, is likely to be folly. There is good reason to question what motives privately-funded people, let alone state organisations, would have for exploring the far reaches of space anyway. So, every incentive must be found to drive groups to establish enterprises for resources, tourism or whatever else is deemed necessary in order to expand.

A Marxian system of equality in outer space, even if this could be achieved through some type of agreed social order, will erode the incentive to establish new enterprises from which trade can flow. Arguments that the driven will explore and settle space to expand the reach of humanity, to make it a multi-planet species and to advance science, independent of any economic incentive, are likely to be as ineffective as they were in stimulating productivity in the communist states of the Earth. People are motivated by these laudable and altruistic arguments, but the impulsion to work to secure the common good is not common and it is probably not sufficiently universal to be sure of achieving the results required in outer space.

The sense of community, which is perceived to be another golden egg of the Marxian vision, depends of course on what sense of community we are talking about. As I have elaborated elsewhere, the sense of community in a centrally-organised society driven to complete equality is likely - and very rapidly in the environment of outer space - to turn into a highly autarchic control structure in which there is certainly an evident community, but where the freedom of the individual is all but destroyed [54]. The society of contented slaves is most likely to emerge in an environment where continuously lethal external conditions give every excuse for control structures to expand into lives with ever more vigour by the process of the tendency of humans to expand their power bases.

The maturation of tyranny will be facilitated because the Marxian vision is a single doctrine vision. How exactly this vision will manifest in the environment of space is unpredictable, but any single doctrine society that seeks to protect centrally planned objectives can never tolerate dissenters. It has been recognised for a long time that it is in the nature of single-doctrine societies to remove countervailing views, either through political dictates, peer pressure or the generation of terror and it follows that, *eo ipso*, few lesser arguments need be entertained against the Marxian vision of an extra-terrestrial society. The ease with which the deadly environment can be turned into the common enemy and used to justify the protection and advancement of a single and inflexible political and economic vision makes any social order that promulgates one - and only one path - to social success dangerous. The details of those parts of a Marxian plan that can succeed and those that cannot, therefore, wither into insignificance in the face of the need to encourage a plurality of ideas about how extra-terrestrial society should be ordered.

8. Tyranny and the New Society

New societies in extra-terrestrial environments, having few historical precedents from which they can draw, amount to a blank social canvas on which any determined person or group of people can paint their vision of society. These early stages have some analogy to the societies established by any group migrating across the Earth to previously uninhabited land. As time progresses, there will be other settlements from which lessons and ideas can be drawn, and instead of an entirely blank beginning, each new society will have certain presuppositions, preconceptions and prejudices derived from the great successes, glories, tragedies and catastrophes that other space settlements have witnessed or experienced. Nevertheless, each new group will have the opportunity for a new beginning, which few nations on the Earth today can truly realise, although the ancient Greek city-states may well approximate to some degree the extra-terrestrial case [55].

It is not a great insight to see that this situation of the 'new society' will allow many lessons from the past to be rectified and many new experiments in social organisation to be tested, some of which might well be very successful. However, consistent with my focus here, the principal concern must be identifying those conditions inherent in the new society that might encourage tyranny. Although it is interesting to pursue a speculative enquiry into the types of beneficial and successful new social experiments and arrangements that outer space might allow, I will leave this avenue of discussion for elsewhere because I will assume that any successful arrangement is not one of concern. If the outcome of an experiment leads to a benign, or beneficent, apparatus of governance, then the identification of details of these arrangements does not demand attention, although of course they are of interest as they might prudently be replicated elsewhere.

A lack of cultural diversity will be caused by the barren environment in which most societies will emerge. This early cultural fallowness will also originate from the spatial isolation of many new settlements. However, independent of these causes, this isolation will exist in any new society showing a lack of historical antecedents from which its perspectives are drawn; it will be an affectation of the freshness of an extra-terrestrial social arrangement.

Cultural inventiveness very quickly emerges in any new society as people pursue interests in art, music and literary expression and there is no doubt that, once a settlement is constructed, individuals will, in their spare time, begin to fill these needs. Their expressions of artistic creativity will rapidly permeate and be transmitted and, like any culture, it will take on forms and nuances that will make it quite identifiable as an art form from this or that lunar, Martian or outer space location. Even the smallest Inuit communities in the frigid wastes of the High Arctic of Earth have distinctive artistic expressions [56].

The manipulation of culture by totalitarian states can be accomplished at any time in a society's history and it does not require a malleable new society to allow this to happen [57], but it must be the case that when a society is extremely isolated, and its culture new, then tyrannous people and authorities, if they exist early on, will have an opportunity to mould cultural development in particular ways to serve themselves. They might, for instance, give particular encouragement to artists that paint, draw or photograph depictions of the authority's magnanimous achievements. They might support writers whose narrative is a paean, or at the very least represents a particularly positive view of the social environment, quite at odds with reality. They might support particular architects who will lend to the design of infrastructure a particular style. If this is the first culture to emerge, then unlike state attempts to manipulate the culture of a preexisting society to meet their ends, which often appear transiently successful, but are in fact bitterly resented by the population, the state might well succeed in establishing customs that, at the very least, will influence all subsequent cultural developments.

To control, or merely define in some broad manner, the first cultural artefacts of a new society in space will be an irresistible conduit for the tyrannically-inclined individual or state to exert their presence in society. Cultural icons, great works and monuments are not merely the instruments of influence in the here and now, they are also some of the greatest legacies of dynastic empires; the pyramids of Egypt provide acceptably persuasive evidence of this truth. There is no possibility of predicting what art, architecture or literary works a despotic state would seize upon to ensure a legacy of cultural influence, but the new society offers an open opportunity for an intellectual expression without the contaminating influence of past achievements. If the new works can be made impressive enough, they will overshadow achievements yet to come. As such, there is something of an incentive to divert resources, including human resources, into impressive projects that will not only awe the population in their own time, but will ensure a legacy that shines immutably for many generations to come. In their turn, these very projects will strengthen the influence and the mastery of resources of the very apparatus of governance that gave rise to them.

Of the cultural instruments so prone to the manipulation of authorities, the exertion of control over the promulgation of religious belief is one that can have a lasting influence over thinking and behaviour, well beyond the time of a settlement's founding. Dynastic institutions, particularly those of monarchy, but by no means these alone, have throughout history sought either to abolish religion or to make it subordinate to their political and social authority, or they have taken quite the opposite route and instilled a belief that the authorities are a direct link to the deity that the people worship.

Some factors will mitigate the chances that religion can be used as an instrument of edification and control in a newly-established society. The intelligence of pioneers selected for the task will provide some measure of resistance to the success of such an enterprise, particularly if they are selected for their secular views. However, the forces that support the abuse of religion are extremely powerful. The paucity of culture and prior social order creates a mental power vacuum in any new settlement, which can be filled with directed and specific religious thoughts by a determined despot. The lethality of the environment, which may be a new experience for early settlers, can be relieved by religious security; an intelligent group of leaders will soon realise that this belief can be controlled by them as a means to secure their own power base and ensure a loyal following by the people [58]. Religion provides a common set of beliefs that can be used for more than the coercion of new arrivals into conformity within the system; if it is overt enough, the religious framework will provide an immediate and obvious jigsaw of social arrangements into which the newcomer can fit.

I should make it clear here that when I refer to 'religion' I make no pretence to predict what exactly this is. It could be a God, gods or even some new form of extra-terrestrial paganism, a type of worship of the extra-terrestrial environment in the face of its overwhelmingly destructive power. It could be something entirely unknowable to us, but the term 'religion' I take merely to be some form of collective worship of an entity outside of the mere practicalities of everyday life.

Perhaps more alarming is the recognition that it requires no iniquity or malevolent intentions for religion to find its way into the decisions and behaviour of authorities; if a majority or a powerful coterie of individuals all share a common belief then, even without their direct efforts, their religious beliefs will begin to pervade the way in which society is run. In relatively small groups of people, at least settlements far smaller than a typical terrestrial nation-state, these influences will be less easily diluted, and will be more likely to make an imprint on the style and manner of governance. A narrow and grey separation exists between minor religious influences, which show up in the activities of those who govern and are an innocent reflection of a religious belief that happens to prevail, and more pervasive religious influences that are deliberately used as an instrument of coercion or just mild compulsion. As the religiously-inclined hold beliefs with sometimes innocent, but completely unshakeable conviction, they sometimes cannot tell the difference between actions that are carried out with a moral certitude, having foundations in specific religious beliefs, and actions that amount to coercion of others. Where the religious belief dovetails with most social expectations, such as the unacceptability of theft, then there is little reason for concern, but where it does not, then there is every chance of religiously inspired mechanisms of tyranny in outer space.

There is no modern society that does not place importance on the ability to track time and there can be little doubt that societies in space will equally require reliable time keeping for planning and organisation in all its varieties. The different lengths of days, months and years on other planetary bodies in our Solar System, or even the lack of them in stationary points in free space, will force upon new settlements novel means of time keeping. Even Earthbound explorers who have attempted to coordinate exploration on other planetary bodies, such as Mars, have had to find ways of adjusting schedules to take into account the different day lengths in those environments [59].

These factors collude to force upon any new space settlement the need for a different calendar compared to the Earth. In outer space, without the confining and defined regularity of the planetary spin and year to define calendars, almost any arrangement could be adopted in any outpost, provided it remains consistent.

Revolutionary political movements on the Earth have seized upon the calendar as a means to establish their authority over a society. What better way to establish pervasively one's preeminent power every day, month and year of a person's life than by renaming those very intervals of time? The French Republican Calendar, which was used from 1793 to 1805, definitively re-assigned almost every interval of the year. Each week had ten days and each of the twelve months had thirty days. Months received names reflecting weather patterns and agricultural motifs (e.g. fog, frost, pasture), the days were named after animals if they ended in five (e.g. horse, goose), after tools if they ended in zero (e.g. plough, vat), and plants and minerals for all others (e.g. celery, pear, coal, bitumen).

There is something vaguely praiseworthy to be said for the fact that the leaders of the French Revolution did not name any of these intervals of time after themselves, but the mere act of changing the calendar is the supreme, fervid demonstration of power over the definition of the unchangeable forces of nature that map our lives, when it is that we were born and when it is that we die.

None of these revolutionary movements lost their opportunity to re-define the very beginning of socially significant time. There is something momentous about changing naming systems, but the definition of the first year of an epoch establishes outright that the beginning of time occurred within the period of the revolutionary governance. Revolutionary France, in a decisive mark of indefatigable Jacobin centralisation, declared Year One to be 1789. Pol Pot (Saloth Sur), whose murderous Khmer Rouge-led regime took the lives of up to two-and-a-half million Cambodians in a drive to establish a utopian agrarian nation, decreed Year Zero as 1975, the year he seized power from Prince Sihanouk.

It is no easy task to revise an entire calendar on the Earth, except in circumstances in which complete power over a nation-state is ensured and a level of despotism is secured within which the inconvenient - and apparently maniacal decision - to revise the calendar, can be implemented with complete acquiescence from the population. In outer space, however, this is an entirely different matter.

The laws of physics provide not merely the excuse, but also the absolute necessity to devise a new calendar in any new location in outer space. For any new state that presides over the organisation of a settlement, the definition of the calendar can be their first, and quite legitimate, demonstration of power. As on the Earth, the establishment of a growing number of settlements on other planetary bodies, particular the moon and Mars, for instance, will require the need for a regulated and agreed upon system of time-keeping, particularly if trade and other activities are to be successfully coordinated. Eventually, this growing conformity in time systems will decrease the ease with which a calendar can be altered without it being an act of revolutionary provocation.

Surely many people must have perceived a certain absurdity when, in the late eighteenth century, French revolutionaries announced their plans to exchange a well-established calendar for one in which the tenth day of the month of Thermidor was to be named 'Watering Can' and the thirteenth day of the month of Bumaire, 'Jerusalem Artichoke'. With any familiar system of nomenclature, its absurdities and anomalies become invisible to a population who have become acquainted with it through generations. Any new system's idiosyncrasies begin by standing out from afar as idiocies of the highest degree, particularly when there is no apparent need to change a well-working system other than to indulge the appetite for change for its own sake.

There is no such indulgence in an environment where nature does not provide such obvious constraints on how a calendar should be organised, where there is no precedent that has run for many millennia, a diversion from which seems quite gratuitous, and where the monotony of the environment and the difficulty of achieving some semblance of collective uniqueness can be partly achieved through one's own distinctive method of measuring time. Under such circumstances, one of the great instruments of political revolution becomes a matter of logical development.

In extra-terrestrial surroundings then, the calendar is just one instrument by which any state or form of governance, but particularly one prone to control, can assert a distinctive and powerful influence over the collective psyche. A degree of permanence of the political and social order is suggested through the definition of the mode of time-keeping and a reminder to the individual that it would be foolish to challenge a system that has defined the very means by which time shall be measured.

The establishment of a new settlement in a lethal extra-terrestrial location will require some surveillance to detect faults in construction and prevent criminality, in the longer term, which could have catastrophic consequences. As the smallest leak in a pressurised system is a potential disaster, there are excuses for the establishment of comprehensive interlinked video and sensor monitoring systems.

As potential weak spots within the physical structure or places where criminal activity might be focused are hardly possible to predict with complete certainty, the new society will suffer from the need for a most intensive and thorough-going surveillance regime.

There are, nevertheless, some special characteristics of the extra-terrestrial case that must be highlighted, which lend themselves especially to the feared use of these instruments for tyrannical conduct. Unlike the Earth, there is no place - no place at all - where those who wish to escape cameras can go. Whilst we may argue with good cause about the impact of video cameras in towns and cities, the state's ultimate reach on the Earth is constrained by our ability to organise revolution in the countryside. We may never choose, or need, to exercise this prerogative, but it exists nevertheless and it provides a final defence against the state (unless by some draconian system of hearing and seeing satellites, even conversations in every house in the depth of the countryside are ultimately monitored, but such a situation would hardly be tolerable). Even a pressurised human rover on another planetary surface can be easily monitoring and conversations within it recorded. Thus, a concern about the emergence of a *bona fide* surveillance state in the initial stages of the establishment of a settlement is a real one. It would be very easy to generate a culture of intimidation based on the knowledge that nowhere is free of surveillance and that this monitoring is legitimately required to ensure the safety of everyone [60].

A liberty-seeking authority can easily engineer a system where defined places are not under surveillance at all, where safety checks are carried out to ensure that continuous surveillance is not required and where rapid reaction airlocks and other devices ensure that these areas can be sealed off in the event of some type of catastrophic criminal action that was not detected until it had rendered damage. The avoidance of a surveillance state would therefore have to be engineered into the architecture of a new settlement, even at its design stage - another instance of how the pursuit of liberty affects the very design of extra-terrestrial outposts [25].

In the early stages of the existence of a settlement, it is a matter of great importance that the free exchange of ideas, conventional and unconventional, is encouraged and facilitated by every physical and social mechanism available. The tendency of the population to succumb to tyrannous authorities and the natural tyranny, towards which the environment will force society, makes a vibrant political state absolutely essential. Now, if the state should feel inclined to prevent this out of its own expedient self-interest, then it will not find it hard to do so, and one means by which it can put the brake on the free exchange of economic and political information is through a surveillance state. From the excuse that the new society requires extensive surveillance to ensure its safety will emerge the systems of monitoring that can be used to quash the spread of information in open spaces and this will rapidly become the norm for the population. From this situation they will find it difficult to imagine any other arrangements, and they may feel more fear from the possibility of the removal of these systems and the impact that this would have on the certainty of safety.

People may not be aware of the impact on their economic situation, for what is lost is not observed, but the stifling of the exchange of information will drastically reduce economic productivity and the efficiency of industries, and it will lead to the conditions for extra-terrestrial economic tyranny discussed earlier.

But there is a much more fundamental contradiction in the extra-terrestrial surveillance state. If the surveillance culture discourages the exchange of information related to nefarious and dangerous cost-cutting and substandard engineering by the authorities or companies with links to the authorities, then the safety benefits gained by pervasive video monitoring may be more than offset by the danger caused by the culture it encourages: one involving a lack of open and free communication.

9. Conclusions

No-one on the Earth is entirely physically free. The oceans or vast tracts of land hem-in every citizen in some way. Nor, in most nation-states, are the vital commodities on which people depend, particularly food and water, delivered without passing through elaborate production and supply networks that no single individual could master and that often operate under the oversight of state regulations.

Why then, should extra-terrestrial tyranny especially worry us? The lack of breathable air in the extra-terrestrial environment and the problems this causes in movement I believe create a fundamental, and categorical, difference in the type of despotism that can emerge in those environments. The psychological repercussions and the social apparatus that can emerge in a society where something that is required on a second-to-second basis for life, and its want can instantly kill (namely oxygen), creates a new instrument of tyranny, the potential and scope of which has not been witnessed before.

Nevertheless, arguments about whether the type of tyranny that might develop in space is something categorically different, or merely a matter of degree, do not affect the reasons why one should consider this to be a topic worthy of special discussion. Tyranny has had appalling consequences on the Earth since the birth of civilisation. It has engulfed and destroyed hundreds of millions of lives and it has transformed once prosperous, civilised and advanced nations into depraved, fanatical and wanton barbarians. Even if one was to somehow imagine that tyranny in space could be no worse than on the Earth, the history of terrestrial societies would give us every reason to examine the conditions for liberty and tyranny in space.

Poised at the top of a gravity well - at least with respect to the Earth - tyranny in outer space will cause an irritation and, eventually, a grave threat to terrestrial liberty. And, from this perspective, it presents a quandary and a social challenge entirely different from terrestrial tyranny. The manner in which this tyranny would eventually manifest, and the exact means by which it would interfere with terrestrial affairs can only be speculation, but preventing despotism in outer space is as much a matter of serious concern to nation-states on the Earth as it is for the communities who eventually struggle directly under it.

There are many types of structure under which societies beyond the Earth might eventually emerge. On the Earth, World Space Agencies and similar global alliances of space exploration have been proposed [61]. All of them will, by definition, emanate from the Earth, and the first settlements to be established will be constructed under the auspices of these organisations or alliances. If these organisations are not to contribute inadvertently or deliberately to the emergence of tyranny, one of their key responsibilities must surely be to ensure that extra-terrestrial communities decide for themselves what their optimal conditions for liberty are to be and how they are to thwart tyranny. As with the history of any relationship between mother country and colony, attempts to fashion liberty along preconceived terrestrial lines will almost certainly contribute eventually to the milieu of dissent, revolution and thereby the conditions necessary for despots to use these conditions to focus enmity.

Eventually, the link between the Earth and other settlements as well as the relationship between those settlements might be governed by some type of League of Worlds, an organisation that recognises that the structures of governance in any particular extra-terrestrial location and the mechanisms needed to nurture liberty, and prevent the inexorable drift to tyranny, must be decided by the people who live under them. Such an organisation would promulgate the general philosophy of liberty-seeking, whilst still allowing each planetary body or place in space to work on the emergence of its own brand of liberty, under its own specific set of challenges. By nurturing links between all settlements and the Earth, it would serve many political and economic purposes. It would encourage, and could actively develop, the conditions for free trade, so vital in preventing economic and political tyranny emerging from autarky. It could reduce the chances for a tyrannous regime infecting others with a culture of coercion by acting as a bulwark, and it could ensure the maintenance of political and economic links between the Earth and developing settlements in space, thus vitiating the paranoia that might develop on the Earth in response to the emergence of a tyranny in space. Such an organisation would act as an interplanetary trading organisation and a focus of political mediation.

Ultimately, forms of extra-terrestrial tyranny may be different from those we have seen on the Earth. But the human character, history has shown, remains invariant and the prediction that tyranny will claw its way into settlements in space, with all the abominations it is capable of bringing forth, will certainly be realised. It is not beyond the inhabitants of the Earth, and the denizens of the realms beyond, to successfully frustrate its progress.

Notes and References

1. A point made by David Hume (e.g. D. Hume, *"Essays: Moral, Political and Literary"*, Liberty Fund, Indianapolis, 1987; first published 1758), although Hume still believed it was possible to accept a version of liberty that, despite its faults, adhered to basic human freedoms, in his case this liberty was that of eighteenth-century Britain.
2. The literature from which these lessons can be drawn is vast, but none of it touches upon the conditions for liberty in outer space, although many science fiction writers have explored issues of social conditions in space (for example, for the Moon specifically, R.A. Heinlein, *The Moon is a Harsh Mistress*, Hodder and Stoughton, 1966, and for Mars, K. Stanley Robinson, *The Mars Trilogy*, Collins, London, 1999).
3. T. Hobbes, *"Leviathan"*, Oxford University Press, Oxford, UK, 1998 (first published in 1651), J. Locke, *"Two Treatises of Government"*, Everyman Library, New York, USA, 1993 (first published in 1689), J.-J. Rousseau, *"The Social Contract"*, Penguin, Harmondsworth, UK, 1976 (first published in 1762).
4. e.g. J. Rawls, "Political Liberalism", Columbia University Press, New York, 1993; R. Nozick, *"Anarchy, State, and Utopia"*, Blackwell, Oxford, 2003.
5. I touch briefly on the justification for the extra-terrestrial state in C.S. Cockell, "Liberty and the Limits to the Extra-terrestrial State", *Journal of the British Interplanetary Society*, **62**, pp. 140-141, 2009.

6. C.S. Cockell, "An Essay on Extra-terrestrial Liberty", *Journal of the British Interplanetary Society*, **61**, 255-275, 2008.
7. I do not intend to imply that society as a whole is on a pre-defined trajectory and that by attempting to unravel some type of 'laws' of history the unavoidable destiny of extra-terrestrial society can be revealed. This would be to admit of the historical and inviolable determinism of terrestrial human history supported by Marx, Hegel and others. I reject this thesis; my view is aligned rather with the view of terrestrial society elaborated by Popper, Herzen, von Mises, Hayek, Berlin and others – that human history can be altered and determined by individual decisions and the collective decisions made by groups of individuals. The only point I claim here is that the extreme physical conditions of outer space make the trajectory of the facets of society that give rise to extra-terrestrial tyranny predictable and that it is therefore in the power of individuals and institutions to pre-empt these sources of despotism and minimise the chances of their emergence and growth.
8. It should equally be recognised that the rooting of democracy in space may also lead to its more effective spread. There are countless examples through human history of the tendency of either tyranny or freedom in one location to infect nearby settlements. The best analogy for the extra-terrestrial case is surely the way in which the relatively isolated city-states of ancient Greece influenced one another. On the weakening of ancient Greece in the third century BC and the effect this had on the city-states, Fowler observed: 'At a time when she was weak and easily broken, dispersed in a variety of independent cities, the Achaeans first united themselves: and then attaching some of the neighbouring cities by assisting them to expel their tyrants, whilst others voluntarily joined them for the sake of that unanimity which they beheld in so constituted a government, they conceived the design of incorporating Peloponnesus into one great power'. Fowler, W.W. *"The City-State of the Greeks and Romans: A Survey, Introductory to the Study of Ancient History"*. Chapter X; Macmillan, Tennessee, 1913. The relationship of ancient Greece to Sparta in earlier times also provides examples of the spread of tyranny through relatively disconnected city-states.
9. Everrett Dolman provides a compelling analysis of the geopolitical factors that will influence power in outer space (E.C. Dolman, *"Astropolitik: Classical Geopolitics in the Space Age"*, Frank Cass, London, 2002). He discusses the way in which gravity wells and other physical characteristics of the space environment will result

in places of greater or lesser political importance with the chilling observation: 'However, the astropolitical dictum that control of certain terrestrial and outer-space locations will provide a distinct advantage in efficiency and will lead the controller to a dominant position in commercial and military power seems assured' (p. 84).

10. L. Dumont, *"Essays on Individualism: Modern Ideology in Anthropological Perspective"*, Chicago University Press, Chicago, 1992, p.40.

11. For example, Lunar Liberty will not be Australian, American, British, or Chinese Liberty, or indeed any other variant. It will be exactly what it says it is: *Lunar* Liberty. Lunar Liberty will be the amalgam of the specific historically contingent factors and social conditions that give rise to notions of what is important in the list of freedoms of lunar settlers. Mixed into this pot will be the specific metaphysical views of individuals exposed to the lunar environment and its resultant sociology. Their view of which aspects of so-called negative and positive liberty are important in the extra-terrestrial environment will ultimately fashion their view of what constitutes "liberty" or "freedom" at its most fundamental level. Despite the necessary humility that is required in the face of this when discussing extra-terrestrial liberty, I think it is possible to identify factors that will lead to generally coercive and repressive conditions with respect to the expression of human creativity and free thinking, and the rejection of which does not threaten the safety of an extra-terrestrial settlement.

12. J. Locke, *'Two Treatises of Government"*, section 57 ('Of Paternal Power'), 1689. Locke made this statement in the context of discussions about the usefulness of laws.

13. J.S. Mill, *"On Liberty"*, Oxford University Press, p.107, 1998.

14. A point made by Bertrand Russell in another arena in which the state can implement widespread and apparently justifiable public safety precautions – health care: 'The most obvious example of a matter where the general welfare depends upon a universal minimum is sanitation and the prevention of infectious diseases …The stamping out of malaria and yellow fever by destroying mosquitoes is perhaps the most striking example of the good which can be done in this way. But when the good is small or doubtful, and the interference with liberty is great, it becomes better to endure a certain amount of preventable disease rather than suffer a scientific tyranny' (B. Russell, *"Principles of Social Reconstruction"*, Routledge, London, p. 49, 1997).

15. E. Burke, *"Reflections on the Revolution in France"*, Penguin Classics, London, p. 205, 1986.

16. There is a distinction to be made here between arbitrary interference and arbitrary domination, a point taken up and explored in depth by Pettit (P. Pettit, *"Republicanism: A Theory of Freedom and Government"*, Oxford University Press, Oxford, 2010). They are not the same. One can be dominated by a state in ways that are not directly interfering and one can be interfered with, without obvious domination. Here I consider tyranny to be actions where an extra-terrestrial state gratuitously interferes *or* dominates, i.e. exerts unnecessary coercive pressure on a population.
17. A point taken up by G. C. MacCallum, "Negative and Positive Freedom", In D. Miller, *"Liberty"*, Oxford University Press, Oxford, p. 100-122, 1991. MacCullum regards freedom as constituting an agent's freedom from restraint in order to do some action. In the context of this relationship, negative and positive freedom are just foci on different parts of it – negative freedom a focus on the lack of restraint, positive freedom a focus on the ability to do certain things once a restraint has been removed. Seen from this perspective, there is no categorical distinction between negative and positive forms of freedom.
18. A perspective eloquently explored by Quentin Skinner (Q. Skinner *"Liberty Before Liberalism"*, Cambridge University Press, Cambridge, 2004). In particular the following captures the possibility that the threat of coercion that might be required to maintain safety in extra-terrestrial environments may be sufficient to constitute despotism: 'The thesis on which the neo-roman writers chiefly insist, however, is that it is never necessary to suffer this kind of overt coercion in order to forfeit your civil liberty. You will also be rendered unfree if you merely fall into a condition of physical subjection or dependence, thereby leaving yourself open to the danger of being forcibly or coercively deprived by your government of your life, liberty, or estates. That is to say, if you live under any form of government that allows for the exercise of prerogative or discretionary powers outside the law, you will already be living as a slave. Your rulers may choose not to exercise these powers, or may exercise them only with the tenderest regard for your individual liberties ... The very fact, however, that your rulers possess such arbitrary powers means that the continued enjoyment of your civil liberty remains at all times dependent on their good will ... And this...is equivalent to living in a condition of servitude' (p. 59-70).
Much of this concern might be mitigated by ensuring that arbitrary powers are minimised, for example by ensuring a rule of

law, but authorities can still threaten the use of ambiguous legal structures and laws against those whom they wish to control.
19. In the words of Mill: 'Like other tyrannies, the tyranny of the majority was at first, and is still vulgarly, held in dread, chiefly as operating through the acts of public authorities ... But ... its means of tyrannizing are not restricted to the acts it may do by the hands of its political functionaries. Society can and does execute its own mandates ... it practices a social tyranny more formidable than many kinds of political oppression, since, though not usually upheld by such extreme penalties, it leaves fewer means of escape, penetrating much more deeply into the details of life, and enslaving the soul itself. Protection, therefore, against the tyranny of the magistrate is not enough: there needs protection also against the tyranny of the prevailing opinion and feeling; against the tendency of society to impose, by other means than civil penalties, its own ideas and practices as rules of conduct on those who dissent from them... There is a limit to the legitimate interference of collective opinion with individual independence: and to find that limit, and maintain it against encroachment, is as indispensable to a good condition of human affairs, as protection against political despotism.' J.S. Mill, "*On Liberty*", Oxford University Press, Oxford, p.9, 1998.
20. Not everyone is convinced that even in a liberal democracy the media operates as freely as it should. Even when the greatest efforts are made to ensure the free exchange of information, many factors, including the influence of large, dominating corporations, can create conditions for the manipulation of information. These views are, however, merely an argument in favour of the need for the freest exchange of information possible, given that even with these efforts, the media is still unlikely to be completely objective. Two penetrating critiques of this problem are made in N. Chomsky, "*Media Control: The Spectacular Achievements of Propaganda*", Seven Stories Press, New York, 2002, and E.S. Herman and N. Chomsky, "*Manufacturing Consent: The Political Economy of the Mass Media*", Bodley Head, London, 2008.
21. Although, of course, we are all subject to certain prejudices that are a product of our history and social condition. No consensus formed in the public sphere is completely objective. But the point must be made that, at the very least, every effort should be undertaken to allow the free exchange of information and the free gathering of information so that, accepting the inevitable contingency of consensus views, they do at least approximate to some degree of objectivity within the unavoidable limitations of the information available at that time.

Subjective views will always mingle with objective assessments if for no other reason that each individual that comprises part of the consensus view is themselves highly biased - sometimes entirely unintentionally: 'Because we generally care more about our own good than the good of others, it is my impression that most people tend to underestimate the negative effects and to overestimate the positive effects of their actions on the good of others' (W.J. Talbott, "*Which Rights Should Be Universal?*", Oxford University Press, Oxford, p. 125, 2005).

22. What constitutes a range of possible options with which one can express one's capacities is another matter for deliberation. As we do not know in what form specific opportunities will emerge in an extra-terrestrial environment, it is difficult to be specific, but the possibility of controlling both trivial options, such as when to go shopping, and options that provide long-term influence in society, are important in creating a variety of possible options through which to express one's freedom to choose. These questions are explored for the terrestrial case by Raz (J. Raz, "*The Morality of Freedom*", Clarendon Paperbacks, Oxford, p. 373-378, 1988).

23. There is a range of policies and regulations. Some are listed here, but the list does not need to be exhaustive to illustrate the regulations under which extra-terrestrial settlements, not yet in existence, are already required to operate. In some cases, however, treaties have not been adopted by some space-faring states, such as the Moon Treaty. All of these regulations and treaties have been developed with entirely positive consequences in mind, but they show the potential difficulties that will emerge in achieving the correct balance between regulations established on the Earth and those fashioned by settlements. Examples are: J.D. Rummel, P.D. Stabekis, D.L. De Vincenzi and J.B. Barengoltz, "Cospar's planetary protection policy: a consolidated draft", *Adv. Space Res.*, **30**, pp.1567-1571, 2002. UN Treaty on Principles Governing the Activities of States in the Exploration and Use of Outer Space, including the Moon and Other Celestial Bodies (1967), Article II, 18 UST 2410, 2413 (1969). Agreement on the Rescue of Astronauts, the Return of Astronauts and the Return of Objects Launched into Outer Space (1968), 19 UST 7570 (1969); the Convention on Registration of Objects Launched into Outer Space (1975), 28 UST 695 (1978); and the Agreement Governing the Activities of States on the Moon and Other Celestial Bodies (1979), 1363 UN Treaty Ser 3 (1984).

24. The problem is clear. State organisations on the Earth, at least in terms of their domestic, national actions, exert an influence

on the people over whom they have legitimate authority, particularly if they are the appendages of a democratically elected government. State space agencies, or global organisations made up of an alliance including these organisations, will ultimately have authority over people who neither operate physically in the places from which these organisations emanate, nor vote for the governments to which the state space agencies belong. Therefore, space agencies, by projecting political power beyond the confines of the Earth, have an inherent tendency to become instruments of arbitrary power, at least as perceived by those people eventually at the receiving end of their decisions.

25. In the essay, "*Liberty and the Limits to the Extra-terrestrial State*", I suggested that another example of this is the modularisation of settlements to prevent logistics, and thus political power, becoming centralised, illustrated in Figure 1. By creating modules with independent systems of supply, it becomes more possible for people, or groups of people, to select different providers for oxygen, water and other vital commodities and to thereby reduce the opportunities for despotic, central oversight. Perhaps this modularisation would also reduce the consolidation of a central system of surveillance by making it easier for concerned citizens to argue that the surveillance of vital machinery and systems is something that they can oversee for themselves in their area, independently of the authorities.

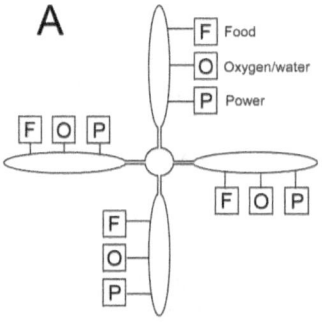

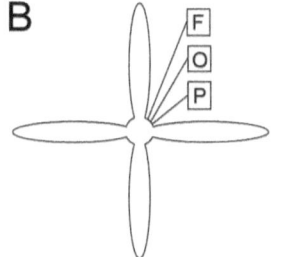

Figure 1. *Extra-terrestrial settlement planning.* The citizens of settlement A have a greater independence of mind and a sense of their liberty than the citizens of settlement B.

Modularisation of the infrastructure prevents all citizens from falling under the control of one authority and it creates a sense of separation which allows small cultural divergences in different areas to develop, leading to greater individualism. Two possible problems that might emerge, however, are: 1) that divergence in the different areas leads to conflict (and possibly complete physical separation of parts of the settlement), which will degrade the sense of community that keeps the whole settlement together, so it remains important that there are common social activities and some facilities that bind all people together; 2) It is easier for an authority to target specific people by attempting to cut off specific resources. In a settlement where oxygen is produced centrally then it is less easy for an authority to cut off a resource that they themselves would need to survive. However, equally they can use the threat of the loss of this resource to manipulate the social situation. As they themselves are affected by the lack of the resource they can legitimately draw upon on a sense of shared community camaraderie in the face of a possible failure and maintain a culture of control throughout the settlement. Dissenters in this type of environment can be made to appear dangerous to the collective good very rapidly. For example, in Figure 1 the people in settlement B might be more inclined to side with the authorities or to believe them because they oversee essential resources.

26. There are endless examples of the stated need for strict and ruthless hierarchies to conduct a war and if the extra-terrestrial environment is regarded as an enemy, then it is obvious that such an ethos can easily be peddled by extra-terrestrial authorities. Here is just one: 'There is no situation in which a body of men are so apt to run into disorder, as in war; where it is impossible that they should co-operate, and preserve the least regularity, unless they are united under a single person, empowered to direct their movements, and to superintend and control their several operations' (J. Millar, *"The Origin of the Distinction of Ranks"*, Liberty Fund, Indianapolis, p. 180, 2006). It is merely necessary to re-read this, whilst replacing 'war' with 'the lethal extra-terrestrial environment' to see how persuasive these arguments become. On the tendency of the state, when under the pressure of war, to lean towards greater interference and control, there is this example from the First World War: 'The government's attitude was that fighting a major war was particularly hard for a free country. Sir John Simon, Attorney General at the beginning of the [First World] war, and later Home Secretary, encapsulated this view: "We

shall be forgiven by posterity if the steps we take are more than adequate, but we shall never be forgiven if the steps we take are not sufficient".' (B. Wilson, "*What Price Liberty?*", Faber and Faber, London, p. 238-239, 2009).

27. A retreat from the constant questioning of a regime is a serious problem and the fact that extra-terrestrial settlements might be established with intelligent people selected with acquisitive and enquiring minds is no guard against the possibility of despotism. For example, von Mises observes: 'The fading of the critical sense is a serious menace to the preservation of our civilization. It makes it easy for quacks to fool the people. It is remarkable that the educated strata are more gullible than the less educated. The most enthusiastic supporters of Marxism, Nazism, and Fascism were the intellectuals, not the boors. The intellectuals were never keen enough to see the manifest contradictions of their creeds ... The absence of criticism makes it possible to tell people that they will be free men in a system of all-round regimentation ... They always tacitly assume that the dictator will do exactly what they themselves want him to do.' (L. Von Mises, "*Bureaucracy*", Liberty Fund, p. 88, 2007).

28. The logic of this is explored by Colin Crouch (C. Crouch, "*The Logic of Collective Action*", Fontana, London, 1982). Membership of a large organisation may not be rational because it costs me to join and my own influence in that large number is miniscule. If the organisation is successful in its objectives (and if those objectives apply across the whole settlement) then I gain the benefit of them if they are good objectives. If they are tyrannical, my membership would not have done much to sway the outcome anyway, so there was no point in taking part. The smaller the population size the more likely my membership will have a perceptible influence and the more likely it is that my view on the topic and my membership will sway others to take the same view. In small extra-terrestrial societies the logic of combining into organisations makes more sense. Although this is good for encouraging involvement in the polity, it may very quickly become an instrument for ostracising those who do not join, or who express countervailing views, particularly as they no longer remain anonymous in a small group.

29. The encouragement of plurality in expressions of religious or social points of view has the effect of mollifying negative criticism not by merely increasing tolerance for a greater set of ideas, but by reducing the latitude to criticise others. The effect of this on the emergence of religious toleration in the United States is explored by Chris Beneke, "*Beyond Toleration: The Religious Origins*

of American Pluralism", Oxford University Press, New York, 2006. Note his comment on page 10: 'In short, as people acquired greater freedom to define their own religious experiences, their liberty to criticize other people's diminished. A pluralistic society required nothing less'.

30. The problem is that an extra-terrestrial society, under conditions of autarky and social isolation, can very rapidly become closed and, as Karl Popper observed concerning closed societies: 'A closed society at its best can be justly compared to an organism ... A closed society resembles a herd or tribe in being a semi-organic unit whose members are held together by semi-biological ties-kinship, living together, sharing common efforts, common dangers, common joys and common distress ... And although such a society may be based on slavery, the presence of slaves need not create a fundamentally different problem from that of domesticated animals'. (K. Popper, *"The Open Society and Its Enemies"*, Routledge, London, p. 17, 1995). His description summarises particularly well the state of affairs with extra-terrestrial societies where common dangers and common distress caused by the external environment are likely to cause a closed and conformist herd instinct to develop.

31. This phenomenon was reported by Toqueville who observed of the French before the Revolution: 'That demeaning form of servitude was always alien to them...They felt for him [the King] both the tenderness one feels for a father and the respect one owes only to God. By submitting to his most arbitrary commands, they were yielding less to constraint than to love; thus they often kept complete freedom of soul even in the most extreme state of dependence'. (A. De Toqueville, *"The Ancien Régime and the Revolution"*, Penguin Books, London, p. 123, 2008). If the authorities secure for people in extra-terrestrial environments complete safety from, and therefore the freedom of not having to worry about, the lethal nature of the outside conditions, then they may well learn to love their authorities even under the worst tyranny.

32. This view has implications for the way in which a society is structured, for it may influence the very design of any constitution established to safeguard liberties. Of the American Constitution: 'They limited the power of the government, but had little to say about the development of the individual. This was because ... American republicans believed that the experience of liberty and self-responsibility would naturally turn people into citizens as they sought to overcome the obstacles inherent in their own

nature and the outside world without being cosseted or coerced by authority' (B. Wilson, "*What Price Liberty?*", Faber and Faber, London, p. 127, 2009). If it is more likely that the extremity of extra-terrestrial conditions will result in negative liberty simply being constricted to an ever smaller sphere, then it may be that Constitutions should say something more than merely the desire to minimise state interference. They may need to say something very forceful about the power of the individual to exercise their capacities within the context of society.

33. This concern with positive liberty was one with which Isaiah Berlin took special issue (e.g. I. Berlin, "*Freedom and its Betrayal: Six Enemies of Human Liberty*", Princeton University Press, 2002). I agree with the Berlin's general view that positive liberty has been abused to convince people that their freedom is best secured under dictatorship and their involvement in a collective will. However, in the sense that encouraging the active expression of individual capacities within a common social framework allows individuals to achieve their own ambitions, rather than merely daydreaming of quixotic projects which they cannot achieve on their own, positive liberty is not always bad. However, the distinction between taking part in a polity so that one's individual ideas, which may be at odds with others, can contribute to society and taking part in a polity to contribute to the agreed social norms and customs, also essential, which contribute towards one's strength in society, is crucial.

34. It is not enough merely to lift restrictions. Once these restrictions are lifted people must be given the potential to realise their ambitions, otherwise their freedom is impotent. As Charles Taylor points out: 'A man's freedom can therefore be hemmed in by internal, motivational obstacles, as well as external ones. A man who is driven by spite to jeopardize his most important relationships, in spite of himself, as it were, or who is prevented by unreasoning fear from taking up the career he truly wants, is not really made more free if one lifts the external obstacles to his venting his spite or acting on his fear' (C. Taylor, "What's Wrong with Negative Liberty?", in D. Miller, "*Liberty*", Oxford University Press, Oxford, 1991, p. 160).

An equally persuasive case for the need to give individuals the appropriate political and economic tools with which to make use of their freedom is made by Amartya Sen (A. Sen, "*Development as Freedom*", Oxford University Press, 1999). Sen also elaborates the importance of democracy as a protective force in this process. It should be noted that most writers who defend the need for this positive sense of liberty vary greatly in the extent to which they support such institutions as the Welfare State, public health provision, etc., all of which are matters of sharp and often divisive discourse amongst liberals (classical and more modern progressive varieties). Nevertheless, the general thesis, which is that freedom is sterile if people are not actively given some means to achieve their desires, is as important for the extra-terrestrial case as it is on the Earth.

35. Although the very struggle of an individual to achieve things within this inherently collectivist environment can itself be liberating, consider the following perspective: 'Habitual exertion is the greatest of all invigorators of character, and restraint and coercion in one form or another is the great stimulus to exertion. If you wish to destroy originality and vigour of character, no way to do so is so sure as to put a high level of comfort easily within reach ... A life made up of danger, vicissitude, and exposure is the sort of life which produced originality and resource'. This passage (p.31) underpins the entire thesis by Stephen, that liberty is only expressed when there is constraint to channel it (J. F. Stephen, "*Liberty, Equality, Fraternity*", Liberty Fund, Indianapolis, 1993). Although Malthus did not claim that restriction is a necessary way to accomplish liberty (that was a later step by Stephen), he did observe a similar point: 'That the difficulties of life contribute to generate talents, every day's experience must convince us. The exertions that men find it necessary to make, in order to support themselves or families, frequently awaken faculties that might have lain for ever dormant; and it has been commonly remarked that new and extraordinary situations generally create minds adequate to grapple with the difficulties in which they are involved' (T.R. Malthus, "*An Essay on the Principle of Population*", Oxford University Press, p. 149, 2008).

In the extra-terrestrial case I have previously pointed out that the inability to move freely in the outside environment may encourage intense creativity in activities that do not require free movements such as some branches of science, ethics and philosophy (C.S. Cockell, "An Essay on Extra-terrestrial Liberty", p. 270).

I prefer to think of the tendency of the extra-terrestrial environment and its repressive nature to be an engine for creativity as a truism, but certainly not an apology or excuse to encourage this repression as a means to invigorate the human mind, as Stephen's view would suggest. Stephen's perspective is to invite any excuse to use legitimately the extra-terrestrial environment to advance tyrannical behaviours and restrictive social practices, under the auspices of the claim that they help bring out extraordinary characters that are necessary for the advance of the settlement.

36. '...but liberty, when men act in bodies, is power' (E. Burke, "*Reflections on The Revolution in France*", Penguin Classics, London, p. 91, 1986).

37. In defending evolutionary development of societies and not revolutionary change, Karl Popper avers: 'In all matters we can only learn by trial and error, by making mistakes and improvements; we can never rely on inspiration, although inspirations may be most valuable as long as they can be checked by experience. Accordingly, it is not reasonable to assume that a complete reconstruction of our social world would lead at once to a workable system' (K. Popper, "*The Open Society and Its Enemies*", p. 167-168). The author goes on to explain that as a new society will always make mistakes, it can never achieve anything better than that which would be achieved by incremental changes. The imperfect new society will always have to be wiped clean and started again. Popper's analysis is quite in agreement with historical experience on the Earth. The problem with extra-terrestrial societies is that they have no prior social order with which to work, except experiences from the Earth, which may be inadequate for the task of building societies in space. So the new extra-terrestrial society will always be prey for zealots and those with utopian visions, as well as facing the constant risk of being wholly reengineered by those with a new revolutionary vision for its development, with the subsequent instability and terror that will result. These ideas for the terrestrial case are also explored by Gray (J. Gray, "*Liberalisms : Essays in Political Philosophy*", Routledge, London, 1991).

    The need for incremental development is emphasized by conservative liberalists such as Burke, Hume, Oakeshott and others who have insisted on the need for moral norms, traditions and customs as the guide for determining what are freedoms and what are not. The requirement they lay down for conservative liberty is, in many ways, the luxury of a society that has a history from which these notions can emerge. In the extra-terrestrial case, although one could still pursue arguments turning on 'natural laws' or 'natural rights', there will be no strong historical social structure upon which to turn.

38. Hannah Arendt makes the observation that, at least in the Greek case, 'the human capacity for political organisation is not only different from but stands in direct opposition to that natural

association whose centre is the home and the family. The rise of the city-state meant that man received "besides his private life a sort of second life, his *bios politicos* ... and there is a sharp distinction in his life between what is his own and what is communal".' (H. Arndt. *"The Human Condition"*, University of Chicago Press, p. 24, 1998). Although this may have been the case in ancient Greece, there is no logical reason why a conflict between family life and political life is inwrought in human behaviour. Although they may be kept quite separate, and people may desire to keep them separate in an extra-terrestrial settlement, provided the authorities respect the privacy of family life, both political involvement and private family life should be entirely compatible.

39. This balance becomes more difficult to achieve when there is greater state interference and a greater number of dictates that make disagreements more likely. On the subject of state regulations and individual objectives: 'All political arrangements, in that they have to bring a variety of discordant interests into unity and harmony, necessarily produce various clashes. From these clashes spring a disproportion between men's desires and their powers; and from these, transgressions. The more active the state is, the greater number of these. If it were possible to make an accurate calculation of the evils which police regulations occasion, and of those that they prevent, the number of the former would, in all cases, exceed that of the latter' (W. Von Humboldt, *"The Limits of State Action"*, Liberty Fund, Indianapolis, p. 81, 1993). If the number of disagreements between individual wishes and those wishes that are more aligned to the state's are to be minimised, it may be better for the former to emerge out of voluntary association rather than more arbitrary state regulations.

40. Many of these individual needs are not necessarily in agreement with others, despite radical ideas that all of our interests can be brought into common agreement. In examining Marxian ideas on the subject of common interests, Brenkert states: 'Indeed, such a unity of (all) interests seems excessively romantic. Even those truly in love may disagree, dispute and differ. What is clearly required is a unity of basic interests as well as (it would seem) agreement on priorities and weightings of those basic interests and willingness to agree on means to realise them' (G. G. Brenkert, *"Political Freedom"*, Routledge, London, p. 125, 1991).

41. Von Mises observed: 'But as a member of society, a man must take into consideration, in everything he does, not only his own immediate advantage, but also the necessity, in every action, of affirming society as such. For the life of the individual in society

is possible only by virtue of social cooperation ... In requiring of the individual that he should take society into consideration in all his actions ... society does not demand that he sacrifice himself to the interests of others'. (L. Von Mises, *"Liberalism"*, Liberty Fund, Indianapolis, p. 14, 2005).
Surprisingly this is in contrast to the views of some modern liberals. Eisenach points out that some progressive liberals, such as Dewey, assert that social progress and personal development are linked because both require 'a unified and articulate will' (E. J. Eisenach, 'Progessivism as a National Narrative in Biblical-Hegelian Time', E. F. Paul, F. D. Miller, J. Paul, *"Liberalism: Old and New"*, Cambridge University Press, Cambridge, p. 65-66, 2007). Despite his reputation to the contrary, even Rousseau, whose famous dictum that individuals can be 'forced to be free' has so appalled people, was clear about his separation of individual wills and the so-called 'general will', which worked to a common unified social purpose and to which he applied his infamous phrase. There is obviously a fine and very important distinction between individual wills, which are respected as being individual, coming together to achieve social purposes as a coherent whole, and those individual wills being artificially (and despotically) moulded at the level of the individual into a common will.

42. Sen observed: 'The loss of freedom in the absence of employment choice and in the tyrannical form of work can itself be a major deprivation' (A. Sen, *"Development as Freedom"*, p. 113).

43. In the smallest settlements, no elaborate formal economic systems, such as banks, will be required for economic activity to occur. Like terrestrial Antarctic stations, exchanges can even be carried out by barter. A particularly famous analysis that is relevant to the extra-terrestrial case is that by Radford (R. A. Radford, "The Economics of a Prisoner of War Camp", *Economica*, **12**, 189-201, 1945). In analysing the emergence of a rudimentary economic system in a Second World War prisoner of war camp, the essential observation of Radford's essay is that in any environment where human beings collect, trade will emerge. In this case, cigarettes became the currency of choice and as Red Cross food parcels came and went the fluctuating availability of items from these parcels, including cigarettes themselves, caused all manner of economic mechanisms, including banking, speculation, attempts at exchange rate control, and so forth, to develop. The analysis shows that in the extra-terrestrial case, no special efforts in constructing economic systems would even need to be implemented for a currency and other economic mechanisms to emerge, although clearly, given that

this is inevitable, formal currencies and banking systems might be established in settlements in anticipation of their need, and to allow for more fluid and formal transactions to be developed.

44. Nevertheless, it might also be observed that the lack of division of labour might also increase the fluid movement of individuals within an extra-terrestrial community between different industries that would indirectly contribute to political freedom and the culture of social mobility that helps sustain it. As Röpke observed in the terrestrial case: 'Thus, as it encroaches on new fields of human activity, the division of labour leads increasingly to mechanization, to monotonous uniformity, to social and spiritual centralization, to the assembly-line production of human beings, to depersonalization, to collectivization – in a word to complete meaninglessness...'. (W. Röpke, "*Economics of the Free Society*", 1963 edition reprinted by the Ludwig Von Mises Institute, Alabama, p.63, 2008).

45. The future of labour relations in space could occupy an entire analysis in itself, but some initial observations on this matter are worth making. In particular, the problem of employee-employer conflict is a matter of importance, particularly in crucial industries such as oxygen production and their allied industries, which produce spare parts and raw materials. As the closure of these industries could threaten an entire settlement, coercive efforts to keep them open encourage economic tyranny. Therefore, understanding labour relations in space directly affects efforts to safeguard liberty. What factors would lead to a breakdown of these relations and even to an extra-terrestrial strike? Figure 2 shows the anatomy of an extra-terrestrial strike, which I derive from the terrestrial case discussed famously by Crouch (C. Crouch, "*The Logic of Collective Action*", Fontana, London, 1982).

The graph shows the price (in arbitrary units) that an employer and employees are willing to pay to end a dispute, against time. For the terrestrial case, curve A depicts the cost to the employer over time after the initiation of a strike. After an initial period of shock after the strike starts, during which an employer makes a calculation to stand against the strike and their resolve strengthens, the cost to an employer increases over time as greater profits are lost due to the strike action. The price they are willing to pay to settle the dispute therefore increases. By contrast, the employees, initially emboldened by their strike action (and they increase their price for settlement as they dig in), soon find their price for agreement reduces as prolonged strike action denies them their livelihood (B). The employees and employer will reach an agreement where lines A and B intersect (point P) at some time after the beginning of the strike. Crouch increases the complexity of the relationship between the curves as he goes on to relate them to the growing loss of profits in the entity concerned. Nevertheless, the basic curves described are sufficient to illustrate key points about the extra-terrestrial case.

Firstly, the price at which employees will settle is likely to be, in the extra-terrestrial case, lower usually than the terrestrial case [B(Et)], particularly for vital industries, as the cost to the employees of the shutdown of production of a commodity required in the extreme environment of outer space is likely to be high. It is likely to engender much greater unpopularity amongst other people dependent on the products of those industries, increasing the peer-pressure to withdraw from such behaviour. The price of agreement for a corporation or the organisation producing the commodity is also likely to increase more quickly than in the terrestrial case [A(Et)^2]. This will be so because of the employer's concern that productivity shutdown will result in loss of confidence in their capacity to produce vital commodities and a possible threat to the settlement as a whole. The price of agreement for the extra-terrestrial authorities that oversee the entire settlement will be even greater [A(Et)^1] because they are likely to be held responsible for allowing production from a vital industry to be curtailed and will be under enormous pressure to bring the action to an end. Thus, extra-terrestrial strikes in vital industries are likely to be very short-lived [P(Et)]. Given these circumstances, is it perhaps unlikely that any vital industry would experience strikes because the cost of the agreement is so great and the effect of a strike so unpopular that an enormous incentive exists for employees and employers to prevent disputes reaching the stage of a strike in the first place. Except in the most trivial non-critical industries, strikes will hardly be a viable mechanism for employees to vent grievances against their employers in most extra-terrestrial environments. This would, on the face of it, considerably reduce employee power in the workplace in the event of a serious grievance against their management. It may even be the case that such strikes would be outlawed. By contrast, though, it could also be the case that the mere threat to strike would have a much greater influence on the employer than would be the case on the Earth, and so in this sense the employees of vital industries could become more powerful. This possibility will depend on the employers' assessment on whether the threat of a strike could realistically be implemented by the employees, and the legal status of strike action in the first place.

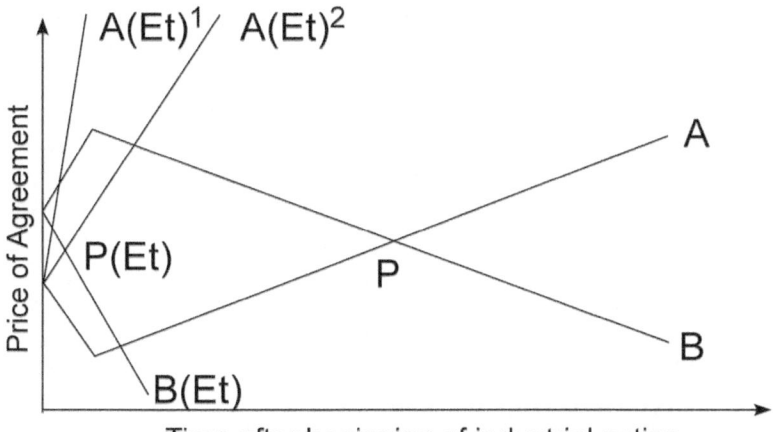

Figure 2. *Economics of an extra-terrestrial strike.* In many extra-terrestrial industries (except the trivial), strike action will be short-lived and unpopular (see text for detail).

46. Hayek uses the case of an oasis to illustrate the problems with monopoly, but his example is unintentionally an outstanding summary of the problem in the extra-terrestrial case if the desert of which he speaks of is simply considered to be the Moon, Mars or anywhere else in outer space: '...if he were, say, the owner of a spring in an oasis. Let us say that another person settled there on the assumption that water would always be available at a reasonable price and then found, perhaps because a second spring dried up, that they had no choice but to do whatever the owner of the spring demanded of them if they were to survive: here would be a clear case of coercion' (F. A. Hayek, "*The Constitution of Liberty*", Henry Regnery, Chicago, p.397–411, 1960). Some have refuted this conclusion with the claim that the spring owner is not coercive by merely demanding the market price for his water, but I think Hayek's point is that the monopolisation of the spring allows the spring owner to coerce the population in matters quite separate to water provision because of his control over a vital resource.

47. The Israeli kibbutz is an example of a small, isolated community that is essentially autarkic. Some people expressed a sense of loss when kibbutzim began to trade with other entities, such as the Israeli state, ending autarky: for example, 'Industrialization, which is a major factor in pulling kibbutzim out of relative autarky ... also offers a strong challenge to interkibbutz equality' (H. Barkai, "The Kibbutz: An experiment in microsocialism" In I. Howe and C. Gershman, "*Israel, The Arabs, and The Middle East*", Bantam,

New York, p. 96, 1972) and 'But it has shown that a socialist form or organisation on the microeconomic and microsocial planes is possible' (p. 97). This rosy view is questionable when on page 87, in examining the availability of goods, the author states: 'By the end of the thirties kibbutzim adopted an allocation mechanism by which each individual was allocated a specified clothing basket composed of several items, which allowed some choice of items within prescribed limits... The clothing and footwear budgets were run much like the point systems perfected during World War II, and cigarettes and toiletries were rationed'. Little more needs to be said. If extra-terrestrial settlements choose the paradise of autarky, then they will likely be living under similar rationing. When economically low output, and the inability to trade, eventually result in the crucial lack of a component essential for survival, for example a part used in an oxygen-producing machine, then the luxury of the kibbutz-like social paradise with its attendant rationing systems may well begin to appear to be an irresponsible indulgence.

48. Trading threatens the strength of established community links in an autarky. Of ancient agrarian societies: 'The basic human instinct to trade would also be disruptive for settled agriculture... Not surprisingly, most of them have looked upon merchants and markets as a necessary evil and sought to suppress them and the market which is their institutional embodiment' (D. Lal, "*Reviving the Invisible Hand*", Princeton University Press, Woodstock, p. 155, 2006).

49. All this amounts to nothing more than the implementation of interplanetary free trade: the enactment of free trade theory. With respect to what happens when a country (or, in this case, as the argument can be easily extrapolated, an extra-terrestrial settlement) pursues autarky, von Mises provides a particularly good summary in his analysis of the effects of autarky on a closed socialist society (L. Von Mises, "*Socialism*", Liberty Fund, Indianapolis, p. 205-220, 1981 and he further develops this analysis in "Economics as Bridge for Interhuman Understanding" in L. Von Mises, "*Economic Freedom and Interventionism*", Liberty Fund, Indianapolis, p. 260-263, 1990).

50. It is often argued that interplanetary trade will be difficult because most products will always be cheaper to produce on the Earth. However, this is to misunderstand basic economic principles. Extra-terrestrial societies need not remain autarkies if their authorities pursue a *laissez-faire* approach to economics such that a Ricardian law of association would operate in trade associations between the Earth and extra-terrestrial settlements, or between extra-terrestrial communities.

To understand the application of Ricardo's insight to the extra-terrestrial case, consider an example involving the Earth and an extra-terrestrial settlement, both of which build small spaceships and air cleaning units. On the Earth, spaceships cost 20 labour days to build and air treatment units 4 labour days to build, whereas in the extra-terrestrial settlement, spaceships cost 42 labour days and air treatment units, 6 labour days to build. In the case of producing both spaceships and air treatment units the Earth has an absolute advantage because both can be produced for less labour time than in the extra-terrestrial settlement. This hypothetical example of the absolute advantage of terrestrial production is the source of the confusion about the viability of extra-terrestrial trade. It represents the fundamental misunderstanding of the principles of comparative advantage. On the Earth, one spaceship can be exchanged for $20/4 = 5$ air treatment units and in the extra-terrestrial settlement one spaceship can be exchanged for $42/6 = 7$ air treatment units. Under autarkic economic conditions, companies on the Earth can exchange one spaceship to get five air treatment units. However, if the Earth exports spaceships to the extra-terrestrial society it can exchange one spaceship and get seven air treatment works instead of just five on the Earth. Thus, provided the advantage gained is not wiped out by the cost of transport back to the Earth, there is a relative trade advantage to be gained by exporting spaceships to the extra-terrestrial community. The trade advantage will also flow the other way. The extra-terrestrial settlement can exchange seven air treatment units and get one spaceship under autarkic economic conditions. However, under open trade with the Earth it needs exchange only five air treatment units to get one spaceship. Under interplanetary free trade, both the Earth and the extra-terrestrial society gain by trading spaceships and air treatment units, despite the absolute advantage enjoyed by the Earth in both commodities. The same would be true between any two extra-terrestrial settlements. Thus, extra-terrestrial authorities need not inevitably become isolated autarkic dictatorships and there may be considerable economic incentive for them to pursue open free trade policies, with effects that will redound to political liberty. The Ricardian "law" of association is particularly likely to work in outer space where communities have a high degree of isolation together with small populations, leading them to specialise in particular products.

51. As stated by Engels: 'in communist society ... Since we know how much, on the average, a person needs, it is easy to calculate how

much is needed by a given number of individuals...it is a trifling matter to regulate production according to needs' (F. Engels, *"Marx-Engels Collected Works"*, Vol IV, International Publishers, New York, p. 246, 1975).

52. The inability to continuously and accurately calculate supply and demand in a socialist society has provided grounds for a serious criticism of socialism, but in a small extra-terrestrial society with access to the information technology not envisaged by Hayek, von Mises and others who have expounded these arguments, an entirely responsive and planned system of production may well be possible. Nevertheless, Marxism in extra-terrestrial environments has greater social problems than merely the technical obstacles to carrying out fast calculations.

53. The need to allow progressive forces to fashion new social policies is particularly great in the extra-terrestrial environment, which will present a new challenge for humans in the combination of its various physical and derived social extremes. Rapid innovation will be the best way to ensure that society can adapt to different conditions found in different extra-terrestrial environments. The problem with a centrally-planned bureaucratic approach is stated by von Mises, in the course of discussing the fixed, dogmatic approach of the Catholic Church: 'The forces that brought about our present civilisation are not dead. If not tied by a rigid system of social organisation, they will go on and bring further improvement. The selective principle according to which the Catholic Church chooses its future chiefs is unswerving devotion to the creed and its dogmas. It does not look for innovators and reformers, for pioneers of new ideas radically opposed to the old ones... No bureaucratic system can achieve anything else. But it is precisely this adamant conservatism that makes bureaucratic methods utterly inadequate for the conduct of social and economic affairs' (L. Von Mises, *"Bureaucracy"*, Liberty Fund, p. 84, 2007).

54. Discussed by Ludwig von Mises: L. Von Mises, *"Socialism"*, Liberty Fund, Indianapolis, p. 169-172, 1981.

55. Peering back through history allows us to attempt to identify a time when the population sizes were still small, but the state of social organisation sufficiently sophisticated to approximate to the extra-terrestrial case. A particularly interesting example may well be the ancient Greek city-states. With populations of thousands and separated by large distances, but nevertheless accomplishing trade, their drift and counter-drift from democracy to oligarchy provides some sense of the way in which extra-terrestrial states might develop independent political characters (Some excellent

classic discussions on the subject are: K. Freeman, "*Greek City-States*", W.W. Norton and Co., New York, 1950; W.W. Fowler, "*The City-State of the Greeks and Romans; Survey, Introductory to the Study of Ancient History*", McMillan and co., London, 1913; J.W. Headlam, "*Election by Lot at Athens*", Cambridge Historical Essays, 1891; A.H.M. Jones, "*Athenian Democracy*", John Hopkins University Press, Baltimore, 1957).

The participatory democracy of Athens is a case in point. Politically built on a succession of legislative developments from those of Solon and Kleisthenes (which might be regarded as analogous to the legislative history that extra-terrestrial settlements will inherit from the Earth), it was sufficiently small and isolated to begin an unparalleled experiment in democracy that continues to awe the world (and encourage a great deal of debate about its effectiveness). Indeed, ancient Athens may not only provide an historical social analogy, but the question should be raised as to whether it might even provide a model for the construction of a democracy in space. The size of terrestrial societies makes it implausible that an Athens-like form of truly participatory democracy can work on the Earth today, but space is quite another matter. A regular meeting of the entire population of a settlement (akin to the Athenian Assembly [*ekklesia*]) and the formation of a Council [*boule*] (perhaps assembled by Lot) to oversee the day-to-day running of the settlement may well be possible.

56. Examples of artistic expression, exhibited by small isolated Inuit settlements, can be found in many books, including: S. Falconer and S. White's "*Stones, Bones and Stitches: Storytelling Through Inuit Art*" (Tundra Books, New York, 2007) and M. Von Finckenstein's, "*Celebrating Inuit Art*" (Key Porter Books, Toronto, 2007). A particularly good historical perspective is provided by E. E. Auger's, "*The Way of Inuit Art: Aesthetics and History In and Beyond the Arctic*" (McFarland and Co. Inc. Jefferson, NC, USA, 2004).

57. There are many examples of how the cult of isolation can develop in nation-states on the Earth. Although it seems implausible to many people that in the twenty-first century, a single individual could wield an uncompromising personality cult over a nation, examples include Turkmenistan, where its leader (Mr. Niyazov) named the days of the week after himself and enforced reading of his autobiography as a requirement for a driver's licence, amongst other characteristically eccentric dictates of geographically and politically isolated leaders. There is every reason to suspect that extra-terrestrial environments, where a single person manages to

gain control of the apparatus of governance, and presides over a population eager for leadership and inspiration, will give rise to similarly bizarre and contorted forms of personality cult.

58. This task becomes easier when the authorities can convince the populace that their religious devotion is inextricably linked with the functions within an extra-terrestrial settlement. Speaking of Puritan values: 'From this reiterated insistence on secular obligations, as imposed by the divine will, it follows that, not withdrawal from the world, but the conscientious discharge of the duties of business, is among the loftiest of religious and moral virtues' (R.H. Tawney, *"Religion and the Rise of Capitalism"*, Penguin Books, Harmondsworth, p. 239, 1975).

59. A wide variety of Martian calendars have been proposed. The great advantage of Mars is that its day length is nearly the same as that on the Earth. The Martian day is only about 2.7% longer than a day on the Earth, on account of the planet's very slightly slower rotation. Thus, a Martian timepiece could be identical to one on the Earth, but running slightly slower, or time can be 'frozen' for 39 minutes at the end of each day, although this would make recording the time of events difficult. Local Martian time was used in the Mars Pathfinder and Mars Exploration Rover missions; watches were calibrated to operate on local Martian time. Decimal time systems for Mars have also been proposed. Although daily time-keeping requires little modification compared to the Earth, this is not the case for the calendar, which must take into account the longer Martian year (668 days or Martian sols). One popular proposition for a calendar is based on the Darian calendar, proposed by Thomas Gangle, which has 24 months. Thomas Gangle describes the longer term corrections to make the calendar work: 'In the Darian calendar, all even numbered years are 668 sols except for those divisible by ten. All other years are 669 sols, so that in ten calendar years there are 6,686 sols. In ten Martian tropical solar years there are 6,685.921 sols, the difference thus being -0.079 sols. A further correction is therefore needed every 100 years, and so every year divisible by 100 is 668 sols instead of 669. With this correction, there are 66,859 sols in 100 calendar years, while there are 66,859.21 sols in 100 tropical solar years. Finally, by making every year that is divisible by 500 a leap year, there are 334,296 sols in 500 calendar years, and the remaining error is only 0.05 sols. Theoretically, this error amounts to only one sol in 10,000 Martian years; however, the actual error will depend on the changes in Mars' orbital elements, rotational period, and the rate of the precession of the pole vector over this period of time.'

This text (taken from the internet at the time of writing) illustrates the fascination that people derive from devising entirely new methods to measure time. This activity is usually the innocent expression of creativity when done by individuals, but when appropriated by states it can become a powerful means through which to assert political authority.

60. As on the Earth, one should not assume that this surveillance is unobtrusive if you are innocent, because it provides extra-terrestrial authorities with the potential to exercise power against anyone, even if, at any particular point in time, they choose not to. It brings us back to the problem of liberty being restricted by the mere fear of the use of arbitrary power and the practical mechanisms that make it possible. For example, on the subject of the use of surveillance equipment to find criminals, which the hypothetical surveillance of substandard engineering works in the extra-terrestrial settlement could easily turn into: 'The assumption behind the 'if you have nothing to hide' claim is that the authorities will always be benign; will always reliably identify and interfere with genuinely bad people only; will never find themselves engaging in 'mission creep', with more and more uses which they can put their new powers and capabilities to, will not redefine crimes' (p.15). It is also based on the belief that 'the manufacturers of ... surveillance equipment have importuned governments into violating the limits of citizen privacy ... by selling them the idea that...the tiny percentage of bad folk among them can be caught' (A.C. Grayling, *"Liberty in the Age of Terror"*, Bloomsbury, London, p. 142, 2009).

61. The following are some of the peer-reviewed examples of these proposals: I.A. Crawford, "On the formation of a global space agency", *Spaceflight* **23**, pp 316-317, 1981; K. S. Pedersen, "Is it time to create a World Space Agency?", *Space Policy* **9**, pp. 89-94, 1993; J. Volosin, G. Kirkham, and L. Guerra, "The dawn of a new space age: developing a global exploration strategy", *AIAA 57th International Astronautical Congress*, **2**, pp. 1284-1293, 2006; G. Gibbs and G. Kirkham, "The global exploration strategy: developing a framework for international coordination and cooperation", *International Astronautical Federation – 58th International Astronautical Congress*, **5**, pp. 3147-3158, 2007.

Other books by Charles Cockell:

Introduction to the Earth-Life System (with Corfield R, Edwards N, Harris N), Cambridge University Press, 2010

Space on Earth: Saving Our World by Seeking Others, MacMillan, 2006

Biological Effects Associated with Impact Events, (with Koeberl C and Gilmour I) Springer, 2006

Project Boreas: A Station for the Martian Geographic North Pole, British Interplanetary Society, 2006

Impossible Extinction: Natural Catastrophes and the Supremacy of the Microbial World, Cambridge University Press, 2003

Martian Expedition Planning, American Astronautical Association, 2004

Ecosystems, Evolution and Ultraviolet Radiation, (with Blaustein AR) Springer, 2003

www.ingramcontent.com/pod-product-compliance
Lightning Source LLC
Chambersburg PA
CBHW020236170426

43202CB00008B/103